Essential Chemistry

States of Matter
&
Phase Equilibria

4th edition

4   3   2   1

ISBN-13: 979-8-8855705-7-2

Sterling Education materials are available at quantity discounts.
Contact info@sterling–prep.com

Sterling Education
6 Liberty Square #11
Boston, MA 02109

© 2026 Sterling Education

Published by Sterling Education

 Printed in the U.S.A.

STERLING
Education

From the foundations of chemical reactions to the complex mechanisms of atomic particles, *Essential Chemistry Self-Teaching Guides* are a comprehensive compendium of clearly explained texts to learn and master these multifaceted chemistry topics.

These guides provide a detailed review of the fundamental mechanisms of chemical and physical processes at the atomic level. Develop a better understanding of the electronic structure of elements, principles of chemical bonding, phases of matter, types and mechanisms of chemical reactions, and principles of solutions and acid-base equilibria. Learn about rate processes in chemical reactions, empirical and molecular formulas, enthalpy, entropy, oxidation number, the laws of thermodynamics, and electrochemistry. Reinforce your learning by working through the practice questions and step-by-step solutions.

Created by highly qualified chemistry instructors, researchers, and education specialists, these books empower readers by helping them increase their understanding of general chemistry.

We sincerely hope that these guides are valuable for your learning.

250909akp

Essential Physics Self-Teaching Guides

Kinematics and Dynamics

Equilibrium and Momentum

Force, Motion and Gravitation

Work and Energy

Fluids and Solids

Waves and Periodic Motion

Light and Optics

Sound

Electrostatics and Electromagnetism

Electric Circuits

Thermal Physics

Atomic and Nuclear Physics

Visit our Amazon store

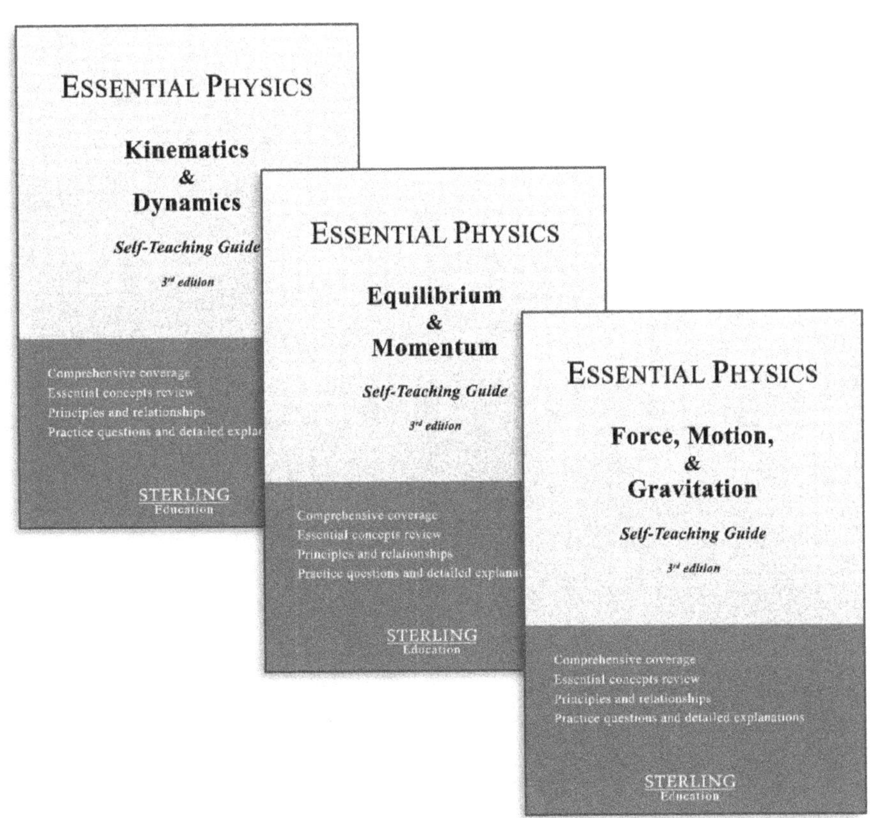

For online practice resources visit

https://www.sterling-prep.com

If you benefited from this book, please leave a review on Amazon so others
can learn from your input. Reviews help us understand our customers'
needs and experiences while keeping our commitment to quality.

Table of Contents

Table of Contents (*continued*)

REVIEW: States of Matter & Phase Equilibria (*continued*)

Table of Contents (*continued*)

REVIEW: States of Matter & Phase Equilibria (*continued*)

Page intentionally left blank

Essential Chemistry: States of Matter & Phase Equilibria

REVIEW

States of Matter
&
Phase Equilibria

Page intentionally left blank

The Nature of Science

Scientific discovery

Observation is the first step toward scientific discovery. Observations recognize patterns and anomalies in unexplained phenomena.

Hypothesis is advanced as a proposed explanation for the observation.

Evaluating a hypothesis involves experiments and additional observations (i.e., recording and collecting data).

Experimental observations are made while manipulating *independent variables* (e.g., how long the plant is exposed to sunlight per twenty-four hours) with *dependent variables* (data collected regarding height measurements).

Observations generate *data* that:

> 1) *support the hypothesis* (i.e., different from proves) or

> 2) *refute the hypothesis.*

Hypothesis are modified by narrowing or expanding their application to explain additional data. More experiments are performed, and the results either *support* (not prove) or *refute* the hypothesis during this iterative process.

Principles are the initially proposed explanations that are specific and apply to a narrow range of phenomena.

Theories build upon valid hypotheses

Theories are proposed to account for the observations when a hypothesis is valid over a range (e.g., sunlight, temperature, or humidity).

Theories are advanced to explain the phenomena and to predict what will happen under other conditions with different variables.

Further observations determine whether the predictions were accurate.

Research (re-search or *again search*) continues as these many observations support hypotheses and theories.

When theories are accurate over various scenarios and experiments, *Laws* are developed based on these repeatedly replicated results.

Laws as established theories

Laws are brief descriptions of how nature behaves in a broad set of circumstances. They are the consolidation of several theories. Laws are the products of multiple theories tested and proven, culminating in a broad claim regarding those predictions.

Laws are widely applicable explanations. Laws with broad and straightforward claims are more robust than those that make claims on a narrower, more specific set of variables and parameters.

The number of Laws describing physical phenomena is minuscule compared to the number of theories and magnitudes, which are less than the number of supported hypotheses.

The number of refuted (unsupported) hypotheses may be several magnitudes larger.

Failed experiments do not typically refer to the physical process where the researcher made a human error during the experiment. A failed experiment means that the *data does not support the hypothesis.* Therefore, after additional trials to validate the "failed" results, the hypothesis must be modified to predict the results of either future experiments or objective observations in the physical world.

Theories, models and laws

Theories emerge from a hypothesis that experimental data has repeatedly supported. Theories are detailed statements that provide testable predictions of the behavior of natural phenomena.

Theories are established to explain observations and then tested (supported or refuted) based on their predictions. Tested theories can articulate the boundaries of the tested hypothesis.

Theories are more specific than *hypotheses* but vaguer and more encompassing than *models*.

Models are constructed to explain the observed behavior if a hypothesis accurately predicts a phenomenon. They are more specific in their application than theories.

Data collected to develop the theories provide the conceptual foundation for constructing models. The model creates mental pictures (e.g., complicated weather patterns displayed as visual pictures for the viewer) of the physical phenomena.

Models should be consistent with scientific theories' predictions and support a comprehensive understanding of a phenomenon. Models allow for predictive outcomes under specific theories (e.g., when the wind speed reaches x, then y will occur).

Multiple models are integrated to explain a portion, or the entire scope, of a *theory*.

Understand the limitations of a model and do not apply it dogmatically unless its applicability is evaluated. A model is an underlying representation of a part of a theory; it is not intended to provide a complete picture of every phenomenon under that theory.

Instead, models provide the fundamental aspects of the theory.

Phase Equilibria

Phases of substances

Most substances exist in *solid, liquid,* or *gas* form.

Solids have a definite shape and volume (e.g., ice). The molecules in a solid vibrate about a fixed position, and a solid *cannot be compressed.*

Liquids have a definite volume but assume the shape of their container (e.g., water). The molecules move about but are close and *bound by intermolecular forces.*

Gases have neither a definite volume nor a definite shape; they occupy the volume and shape of their container (e.g., steam or water vapor). The molecules *move independently* and are not held together by intermolecular forces; therefore, a gas is *easily compressible.*

Freezing and melting point

Freezing point is the temperature at which a liquid changes state to solid (i.e., solidification).

Melting point is the temperature at which a solid changes its state and becomes a liquid.

The melting point of a substance is the temperature at which it changes from a solid to a liquid, and it has the same temperature range as the *freezing point*; the reverse reaction occurs at the same temperature.

Theoretically, a substance's melting point and freezing point should be identical.

Experimentally, a difference in the quantities of the respective processes may be observed.

Vaporization

Vaporization is the phase transition at which a liquid becomes a gas.

There are two types of vaporization: *boiling* and *evaporation.*

Boiling point is when the liquid is heated to a temperature (i.e., the kinetic energy of the molecules) where the vapor pressure is high enough for bubbles to form inside the liquid. It is in the same temperature range as the *condensation point.*

As a liquid boils, the temperature remains constant until the liquid has converted into a gas.

Since the boiling point is dependent on pressure, its precise value varies according to the environment (e.g., elevation relative to sea level).

For example, at high altitudes and low pressure, a substance's boiling point is lower, requiring less heat energy to increase the vaporizing molecule's kinetic energy to break intermolecular attraction to adjacent molecules.

Evaporation occurs at temperatures below the boiling point. It occurs at the liquid's surface when the gaseous phase above the liquid is not saturated with the evaporating substance.

Boiling occurs from the bulk of the liquid, not just the surface.

Evaporation tends to occur more quickly in liquids with *higher vapor pressure.*

Condensation and sublimation

Condensation (the reverse of evaporation) is the process by which a substance changes from the gas phase to the liquid phase. Condensation commonly occurs when a vapor is cooled or compressed to its saturation limit.

*Condensation po*int is the temperature at which condensation begins to occur.

For example, condensation is the process by which water droplets form in atmospheric clouds.

Sublimation (the reverse of deposition) is a phase transition from a solid to a gas, essentially evaporation that occurs directly from the solid phase below its melting point.

Deposition (reverse of sublimation) is the phase transition from a solid directly into a gas (i.e., the matter is not present as an intermediate liquid). This process is observed in the winter when the temperature rises above freezing, and the snow does not thaw into liquid, yet it still melts.

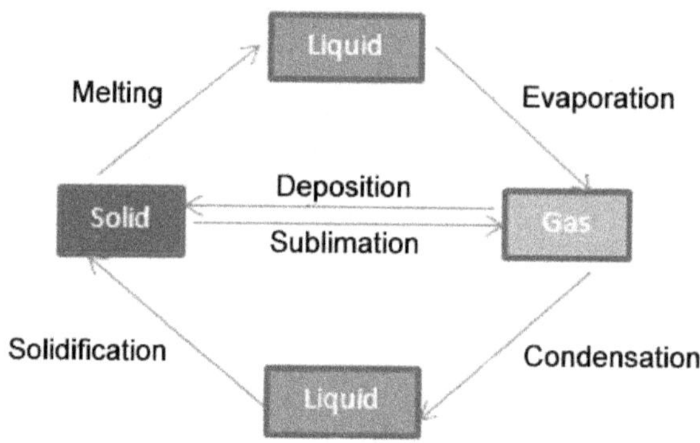

Interconversion of states of matter

Phase Diagrams

Phase change diagram

When an ice block is placed in the Sun, it slowly absorbs the thermal energy through radiation and undergoes a series of phase changes. The ice cube changes from a solid to a liquid to a gas.

The *melting point* of a substance is the temperature at which it changes from a solid to a liquid, and it has the same temperature range as the *freezing point*; the reverse reaction occurs at the same temperature.

The *boiling point* (vaporization) of a substance is the temperature at which it changes phase from a liquid to a gas, and it is the same temperature range as the *condensation point*.

From the graph below, once the ice cube absorbs enough heat to melt (i.e., reaches 0 °C), its temperature remains constant.

Melting is point B to C on the graph, where molecules have reached a temperature of 0 °C.

Ice melts only when this occurs, as shown by the graph's horizontal plateau from B to C.

When liquid water changes to a gas, it absorbs enough heat to vaporize (i.e., at 100 °C). The liquid temperature remains constant from point D to E until all molecules reach a temperature of 100 °C.

The horizontal plateau from D to E signifies the phase transition of vaporization.

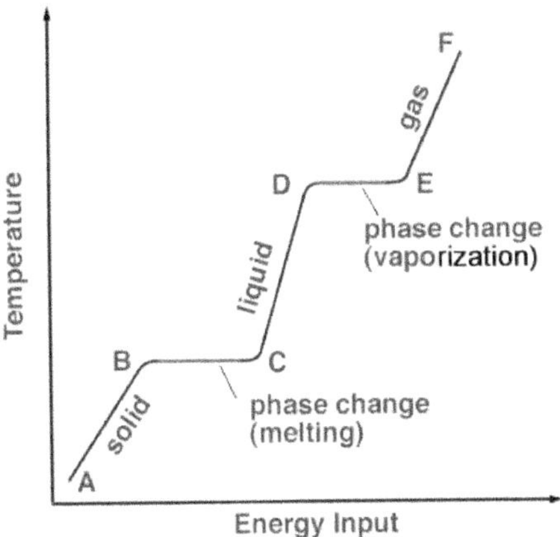

Heat during phase changes

Although the energy (heat) input increases with temperature, it does not remain constant, unlike the temperature during a phase change.

Technically, heat converts the potential energy (*PE*) stored within each atom into kinetic energy (*KE*) before a phase change occurs.

As energy is continuously added to the process, the substance uses the added heat to transition each molecule before increasing the temperature.

Pressure *vs.* temperature relationship

Phase diagrams display the phases of a substance. A phase diagram is a graph plotted as a function of pressure versus temperature.

The graph is divided into three unequal parts, representing the substance's three phases.

Each point on the graph represents a combination of pressure and temperature for the system.

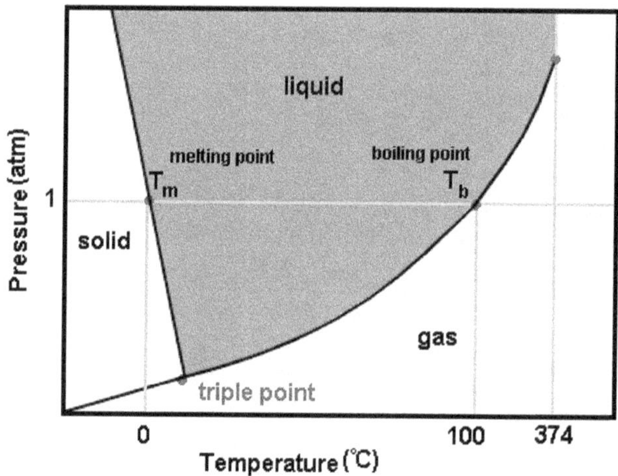

A horizontal line intersects the phase diagram at the temperature at which the substance changes phases.

Melting point and *boiling point* are determined by drawing a horizontal line at 1 atm of pressure, which corresponds to atmospheric pressure.

Triple point is shown when all three lines intersect, and three phases exist simultaneously.

Critical point is the temperature above which a substance exists as a gas, regardless of the pressure.

Molality

Molal concentration

Molality (or *molal concentration*) measures the moles (i.e., the concentration of solutes) in a solvent in terms of the mass (in kg) of the solvent.

The SI unit for molality is mol/kg.

Molality *m* is different from the uppercase *M* for molarity (i.e., *M* = moles/liter).

> *Molality* = (moles of solute) / (mass in kg of solvent)

Calculating molality

For example, if 3.00 moles of sucrose are mixed into 1.00 L of water until the sucrose is entirely dissolved (i.e., saturated solution). What is the molality of the solution? (Use the density of water = 1.00 g/mL or 1.00 kg/L).

Solution: The mass of 1 liter of water is 1.00 kg.

Use the solvent mass of 1. Kg/L and solve for molality.

> *m* = 3.00 mol / 1.00 kg

> *m* = 3.00 mol/kg of sucrose

> *m* = 3.00

Solution: Molality of the solution is 3.00 m.

Instead of expressing the number of moles, the solute's mass is given in grams.

When calculating molality, solvent's mass is the pure solvent, excluding mass of the solute.

Molarity is the volume of solution, including the mass of the solute *and* the solvent.

> *Molality m* is expressed as mol solute/kg solvent.

> *Molarity M* is expressed as mol solute/L solution.

For example, suppose 16.88 grams of sodium chloride (NaCl) are dissolved into 4.00 kg of water. What is the molality of the solution?

Convert grams of NaCl to moles (molar mass of NaCl = 58.44 g/mol).

> moles = (mass) / (molar mass)

> moles = (116.88 g) / (58.44 g/mol)

> moles = 2.00 mol NaCl

Solve for molality:

m = (moles of solute) / (mass in kg of solvent)

m = (2.00 mol) / (4.00 kg of solvent)

m = 0.500 mol/kg of the solution

m = 0.500

Solution: Molality of the solution is 0.500 mol/kg.

Colligative Properties

Solute particles

Colligative properties depend on the ratio of solute particles to solvent but *not* on the type of chemical species.

Solute particles in mixtures increase the strength of intermolecular (e.g., dipole-dipole) bonds. It is more difficult to boil the solution (i.e., boiling point elevation) or freeze the solution (i.e., freezing point depression).

Colligative properties include:

1) relative lowering of vapor pressure,

2) elevation of boiling point,

3) depression of freezing point, and

4) osmotic pressure.

Van't Hoff factor

Colligative properties calculations incorporate the Van't Hoff factor (i), which measures a solute's effect on the solution's colligative properties.

The Van't Hoff factor is:

$$\frac{[\text{concentration of particles produced when a substance dissolves}]}{[\text{concentration of a substance calculated from its mass}]}$$

Concentration (represented in [ ]) is converted to reflect the total number of particles in the solution. (e.g., 5 g of NaCl dissociates into Na^+ and Cl^-).

For example, glucose ($C_6H_{12}O_6$) has an i value of 1 because it does not dissociate in solution.

Nonelectrolytes do not dissociate in solutions (e.g., CH_4 and O_2).

The salt NaCl has an i value of 2 because it dissociates into 1 Na^+ and 1 Cl^- (2 resulting ions per one reactant molecule).

Vapor pressure depression (Raoult's Law)

In 1888, French chemist Francois-Marie Raoult (1830-1901) proved that the vapor pressure of a solution equals the mole fraction of the solvent times the vapor pressure of the pure solvent.

Raoult's Law is expressed by:

$$P = \chi_{solvent} \cdot P^{\circ}_{solvent}$$

where P is the vapor pressure, $\chi_{solvent}$ is the mole fraction of solvent (moles of solvent / total moles of solute *and* solvent), and $P^{\circ}_{solvent}$ is the vapor pressure of the pure solvent.

Calculate χ_{solute} using Van't Hoff factor (i.e., 1 mol NaCl yields 2 mol of ions in the solution).

For example, suppose 2.00 moles of sucrose is added to a pitcher containing 2.00 liters of water. What is the vapor pressure of the sucrose solution? (Use the value 1.00 L water = 1.00 kg, the vapor pressure of pure water = 23.8 mmHg and the molecular mass of H_2O = 18.02 g/mol)

Convert liters of water into mass:

2.00 L water = 2 Kg

2 Kg = 2,000 g

Convert mass of water into moles:

molecular mass = mass / moles

moles = mass / molecular mass

moles = (2,000 g) / (18.02 g/mol)

moles = 110.9 moles H_2O

Solve for the mole fraction, $\chi_{solvent}$:

$\chi_{solvent}$ = (moles of solvent) / (moles of the solute *and* solvent)

$\chi_{solvent}$ = (110.9 moles H_2O) / (moles of the solute *and* solvent)

$\chi_{solvent}$ = 110.9 moles H_2O / (110.9 moles solvent + 2 moles solute)

$\chi_{solvent}$ = 0.98

Raoult's Law to determine the pressure:

$P = \chi_{solvent} \cdot P^{\circ}_{solvent}$

P = (0.98)·(23.8 mmHg)

P = 23.4 mmHg

Solution: P = 23.4 mmHg (compared to initial P = 23.8 mmHg). The addition of a solute caused the vapor pressure to decrease.

Boiling point elevation ($\Delta T_b = K_b m$)

Adding a solute to a solvent stabilizes the solvent in the liquid phase, thus lowering the solvent molecules' tendency to move to the gas or solid phases. Therefore, the boiling point increases.

Boiling point elevation is proportional to the decrease in vapor pressure in a dilute solution:

$$\Delta T_b = k_b \cdot m \cdot i$$

where ΔT_b is the increase in boiling point, k_b is the molal boiling point constant (a given value determined experimentally), m is molality (mol solute/kg solvent), and i is Van't Hoff factor.

For example, what is the boiling point elevation if 6.4 g of ammonia is dissolved in 0.3 kg of water? (Use the k_b for water = 0.52 °C/m and the molar mass of ammonia = 17.031 g/mol)

Convert mass of ammonia (NH_3) into moles:

molecular mass = mass / moles

moles = mass / molecular mass

moles = (6.4 g) / (17.031 g/mol)

moles = 0.38 mol NH_3

Calculate molality of the solution:

molality = (moles of solute) / (mass (in kg) of solvent)

m = (0.38 mol NH_3) / (0.3 kg of water)

m = 1.27 m

Calculate boiling point elevation:

$$\Delta T_b = k_b \cdot m \cdot i$$

$$\Delta T_b = (0.52 \text{ °C/m}) \cdot (1.27 \text{ m}) \cdot (1)$$

$$\Delta T_b = 0.66 \text{ °C}$$

Solution: The boiling point increased by 0.66 °C due to the addition of 6.4 g of NH_3.

Freezing point depression ($\Delta T_f = K_f m$)

Freezing point depression (i.e., lowering the freezing point) is calculated using the following equation: (note that the negative sign indicates a decrease in the change).

Freezing point depression is expressed as:

$$\Delta T_f = -k_f \cdot m \cdot i$$

where ΔT_f is the decrease in freezing point, k_f is the molal freezing point constant (a given value determined experimentally), m is the molality (mol solute/kg solvent), and i is the Van't Hoff factor (given since it is determined experimentally)

For example, a 48.0 g sample of a non-electrolyte (which does not dissociate in solution) is dissolved in 500.0 g of water to produce a solution with a freezing point of -3.5 °C. What is the molar mass of the compound? (Use k_f of water = 1.86 °C/m, the freezing point of pure water = 0 °C)

Freezing point of pure water = 0 °C, and the freezing point depression, $\Delta T_f = 3.5$ °C.

Calculate the moles using the freezing point depression expression:

$$\Delta T_f = k_f \cdot m \cdot i$$

$$3.5 \text{ °C} = (1.86 \text{ °C/m}) \cdot (x / 0.5 \text{ kg}) \cdot (1)$$

$$3.5 \text{ °C} = (3.72 \text{ °C/m}) \cdot (x)$$

$$3.5 \text{ °C} / (3.72 \text{ °C/m}) = x$$

$$x = 0.94 \text{ mol nonelectrolyte}$$

Calculate molar mass by dividing the sample's mass by the number of moles in the sample.

Molar mass = mass / moles

Molar mass = 48.0 g / 0.94 mol

Molar mass = 51.1 g/mol

Solution: The molar mass of the unknown nonelectrolyte is 51.1 g/mol.

Osmotic pressure

Osmotic pressure is the minimum pressure that needs to be applied to a solution to prevent the inward flow of water across a semi-permeable membrane.

Semi-permeable membranes allow solvent particles to pass, not solute particles.

Osmotic pressure (Π) is:

$$\Pi = MRT{\cdot}i$$

where Π is the osmotic pressure, M is the molarity (mol/L), R is the ideal gas constant (0.08206 L·atm/mol·K), and T is the temperature (K)

Osmotic pressure determines whether and in what direction osmosis occurs.

Osmosis is a solvent's movement across a semi-permeable membrane from an area of low solute concentration (i.e., high solvent concentration) to an area of high solute concentration (i.e., low solvent concentration).

Solvent moves from a low Π value to an area across the semi-permeable membrane with a high Π value.

For example, what is the osmotic pressure of a solution prepared by adding 10.5 g of sucrose ($C_{12}H_{22}O_{11}$) to enough water to make 300.0 mL of solution at 25 °C? (Use the molecular mass of sucrose = 342.0 g/mol and the conversion of 300.0 mL = 0.30 L)

Calculate the number of moles of $C_{12}H_{22}O_{11}$ by dividing the mass of the sample by the molar mass of the sample.

moles = (mass) / (molar mass)

moles = (10.5 g $C_{12}H_{22}O_{11}$) / (342.0 g/mol $C_{12}H_{22}O_{11}$)

moles = 0.03 mol $C_{12}H_{22}O_{11}$

Calculate the molarity, M:

Molarity = (moles solute) / (liters of solution)

Molarity = (0.03 mol $C_{12}H_{22}O_{11}$) / (0.30 L solution)

Molarity = 0.10 mol/L

Convert temperature to Kelvin:

$T = {}°C + 273$

$T = 25 + 273$

$T = 298$ K

Since sucrose does not dissociate, calculate osmotic pressure (Π) using $i = 1$.

$$\Pi = MRT \cdot i$$

$$\Pi = (0.10 \text{ mol/L}) \cdot (0.08206 \text{ L} \cdot \text{atm/mol} \cdot \text{K}) \cdot (298 \text{ K}) \cdot (1)$$

$$\Pi = 2.45 \text{ atm}$$

Solution: Osmotic pressure for the sucrose solution equals 2.45 atm.

Colloids

Solution is a homogeneous mixture that consists of one phase (i.e., the solution stays mixed).

Colloid is a solution with microscopic insoluble particles dispersed throughout the solution. A colloid has a dispersed phase (i.e., suspended particles) and a continuous phase (i.e., a medium of suspension that holds the particles).

Colloids stay mixed (i.e., particles will not settle) unless centrifuged under influence of gravity.

A typical colloid is homogenized milk, consisting of butterfat globules dispersed within a water-based solution.

For example, a colloid forms when water and oil are shaken together.

Henry's Law

In 1803, English chemist William Henry (1774-1836) showed that at a constant temperature, the solubility of a gas in a liquid is directly proportional to the gas's partial pressure above the liquid (when the gas is in equilibrium with the liquid).

Henry's Law is described by:

$$P_{solute} = k_H \cdot c$$

where P_{solute} is the partial pressure of the solute at the solution's surface, k_H is Henry's Law constant (unique for each solute-solvent pair), and c is the concentration of the dissolved gas

For example, how many grams of carbon dioxide (CO_2) gas is dissolved in a 0.5 L bottle of carbonated water if the manufacturer uses a pressure of 2.5 atm in the bottling process? (Use the K_H of CO_2 in water = 29.76 atm/(mol/L) and molar mass of CO_2 = 44 g/mol)).

Determine the concentration of CO_2 using Henry's Law:

$$P_{solute} = k_H \cdot c$$

$$2.7 \text{ atm} = [29.76 \text{ atm/(mol/L)}] \cdot c$$

$$c = 0.09 \text{ mol/L}$$

Calculate the moles of CO_2 in 0.5 L by dividing the mol/L by 2.

> moles / grams = molar mass

> 0.09 / 2 = 0.045 mol

Convert moles to grams using the molar mass of CO_2, 44 g/mol (from the periodic table).

> 0.045 mol × (44 g/mol) = 1.98 g

Solution: There are 1.98 grams of CO_2 in a 0.5 L bottle of carbonated water.

Notes for active learning

Gas Phase

Gaseous state

Gas is a fundamental state of matter, distinguished by vast separation of individual gas particles.

Unlike solids and liquids, gases have *indefinite shapes and volumes* because they expand to fill their containers.

As a result, gases are subject to changes in pressure, volume, and temperature.

Some common elements and compounds exist in the gaseous state under normal pressure and temperature conditions, such as oxygen (O_2), nitrogen (N_2), and carbon dioxide (CO_2).

Vapor is a gaseous form usually existing as a liquid or solid at atmospheric pressures and typical temperatures (e.g., water vapor).

Four variables are needed to describe the gaseous state:

1. Quantity of gas, n (in moles)

2. Temperature of the gas, T (in Kelvin)

3. Volume of gas, V (in liters)

4. Pressure of gas, P (in kPa)

Absolute temperature

Temperature is an objective measure of particles' kinetic energy, and the volume of a gas (i.e., the space that it occupies) is proportional to its temperature.

In the 19th century, experiments revealed that gases expand as temperature rises and compress when temperature drops, which led to the determination of the expansion coefficient of gases per degree Celsius.

This posed an interesting question: what happens if the temperature is low enough to reach a point where the gas' calculated volume is zero?

Absolute zero temperature was calculated to be $-273.15\ °C$.

Gases do not reach zero volume because gases liquefy or solidify before reaching this temperature.

Additionally, a gas cannot have a zero volume because the atoms of its molecules occupy a certain amount of space.

Kelvin scale

In 1848, British physicist Lord William Thomson Kelvin (1824-1907) proposed a temperature scale based on this absolute zero value.

Kelvin (K) scale starts with absolute zero as 0 K, and the increments are identical to the Celsius scale.

To convert from Kelvin to Celsius, add 273.15; and for the opposite conversion, subtract 273.15.

Kelvin should be converted to Celsius before converting it into Fahrenheit.

Kelvin temperatures are not written with a degree symbol (273 K *vs.* 273 °C or 273 °F).

Common temperatures expressed in K, °C and °F

	K	°C	°F
Absolute zero	0	–273	–460
Freezing point of water / melting point of ice	273	0	32
Room temperature	298	25	77
Body temperature	310	37	99
Boiling point of water / condensation of steam	373	100	212

Gas Pressure

Pressure as force

Pressure is the average force applied to or experienced by a given area.

Gas molecules are in motion, so in a container, the gas molecules continuously move around and collide with the container's walls. Since the force is applied over a specific area, this force is the container's gas pressure.

Newton's Second Law, $F = ma$, can be used to determine force, which is equivalent to the units of pressure.

The SI units of mass (m) and acceleration (a) are the kilogram (kg) and the meter per second squared (m/s^2).

The unit of force (F) is derived from the equation:

$$F = ma, \text{ is kg·m/s}^2$$

This derived unit is Newton (N).

Pressure is the force exerted over a specific area.

The SI unit of length is a meter (m), so the SI unit for area is m^2.

Therefore, the unit for pressure is N/m^2, known as the Pascal (Pa).

Simple mercury barometer

Barometers measure Earth's atmospheric pressure. 30 inHg is considered normal.

Dry air creates more pressure than wet air, so dry air is more massive than the same amount of wet air.

Consequently, *high barometric readings* (i.e., 30.70 inHg) indicate dry air and fair weather.

Low barometric readings (i.e., 29.80 inHg) indicate an increased chance of rain or showers.

Barometric readings of 27.30 inHg indicate *hurricane conditions*.

Mercury barometers (shown below) are constructed by inverting a column filled with mercury and placing it in a mercury dish, ensuring that no air enters the tube. Some mercury flows out when the tube is inverted, leaving a space in the tube above the mercury. This space is nearly a vacuum. Therefore, there is no pressure pushing down on the column.

The pressure exerted by the mercury column balances the atmospheric pressure. The mercury column's height is about 760 mm above the surface of the mercury in the dish at sea level.

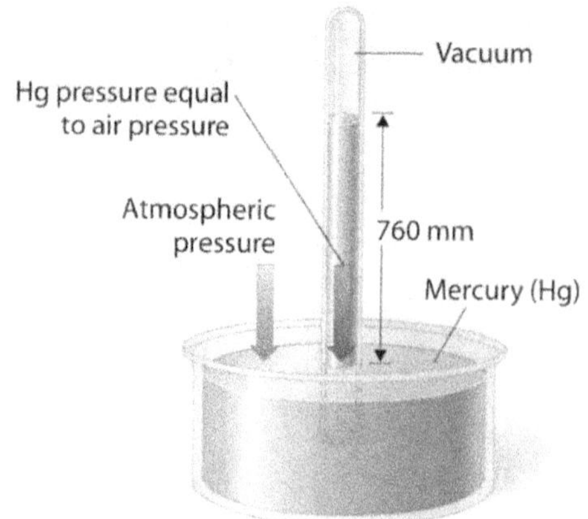

Expressing pressure

Any liquid may be used to measure atmospheric pressure; the height of the column depends on the density of the liquid.

A comparison of mercury (density = 13.6 g/cm^3) and water (density = 1 g/cm^3) indicates that the column would be almost 34 feet high if a barometer were filled with water.

Pressure exerted by the mercury barometer is recorded in units of *millimeters of mercury* (mmHg), a unit sometimes referred to as the torr in honor of the Italian physicist Evangelista Torricelli (1608-1647), who invented the mercury barometer in 1643.

Standard atmosphere (atm) is another unit of pressure, and its conversion factors are listed:

1 atm = 101,325 Pa

1 atm = 760 mm Hg = 760 torr = 76 cm Hg

1 atm = 14.7 lb/in^2 (psi, or pound per square inch)

A bar is a metric unit of pressure equal to 100,000 Pa; it is about equal to atmospheric pressure (101,325 Pa).

Closed-tube and open-tube manometers

Manometer is an instrument used to measure the pressure of gases. It is simply a bent piece of tubing, and in principle, it operates like a barometer.

There are two major types of manometers.

Closed-tube manometer (shown on left below) measures pressures below atmospheric pressure, typically the gas pressure within the container. Since the end is sealed, it contains a vacuum.

Open-tube manometer (shown on right) measures gas pressures near atmospheric pressure.

Pressure is the difference in the heights of the mercury levels in the two arms. The mercury levels in the manometer's two arms relate the gas pressure to the atmospheric pressure.

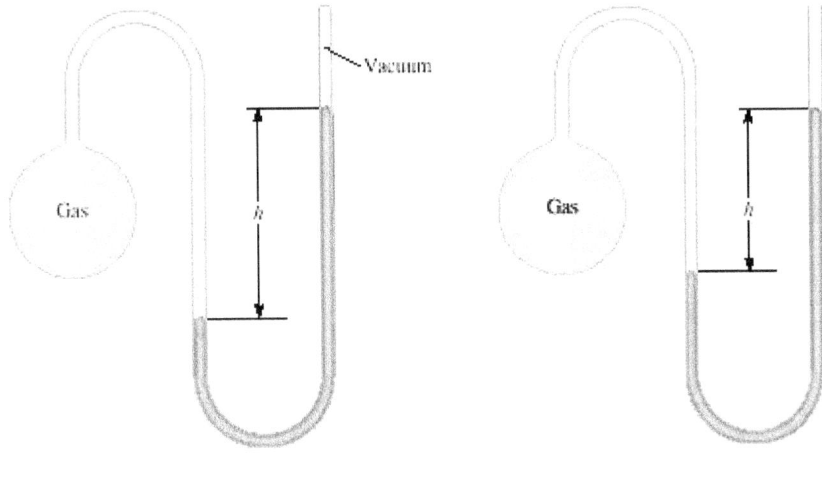

$$P_{gas} = Ph \qquad\qquad P_{gas} = P_h + P_{atm}$$

Left: gas pressure is less than atmospheric pressure (closed tube)

Right: gas pressure is higher than atmospheric pressure (open tube)

Molar volume at STP

In 1811, Italian scientist Amedeo Avogadro suggested that *"equal volumes of gases measured at the same temperature and pressure conditions contain the same number of molecules."*

In chemistry, *molar volume* is typically measured at "standard temperature and pressure" (i.e., STP), which corresponds to a temperature of 273.15 K (0 °C, 32 °F) and an absolute pressure of exactly 10^5 Pa (100 kPa, 1 bar).

At these standard conditions, a mole of gas occupies 22.7 L of space.

Before 1982, when the International Union of Pure and Applied Chemistry (IUPAC, the organization that governs chemistry standards) changed the definition of STP, the standard pressure was defined as 101,325 Pa (1 atm).

Under these conditions, 1 mol of gas occupies 22.4 L.

For problems, use the volume of one mole of gas as 22.7 L, unless otherwise indicated.

Notes for active learning

Kinetic Molecular Theory of Gases

Macroscopic properties of gas

Kinetic Molecular Theory of gases relates the microscopic values of atomic kinetic energy to quantities such as temperature and pressure.

Kinetic Molecular Theory (KMT) utilizes atomic theory to describe the behavior of gases. This theory developed over 100 years, culminating in an 1857 paper by German physicist Rudolf Clausius (1822-1888).

KMT explains the macroscopic properties of a gas (e.g., pressure and temperature) in terms of its microscopic components (e.g., molecules or atoms).

Kinetic Molecular Theory of gases originated in the ancient idea that matter consists of tiny invisible atoms in rapid motion. The idea was developed between the 17th and 19th centuries to explain the properties of gases.

Kinetic theory of gases requires *five critical assumptions*.

1. Gas molecules move in *random molecular motion* and continue moving straight until they collide with another molecule or a surface.

2. Gas molecule is a *"point mass,"* a particle so small that mass is nearly zero.

   An ideal gas particle has a *negligible volume*.

3. *No attractive intermolecular forces* exist for gases.

4. *Collisions* between gas molecules are "perfectly elastic."

   Therefore, the gas molecules' total kinetic energy (KE) remains *constant before and after* the collision since intermolecular attractive and repulsive forces are nonexistent.

5. *Average kinetic energy* (*KE*) for gas particles is proportional to the absolute temperature, regardless of the chemical identity or atomic mass.

   At 0 K (absolute zero or –273 °C), the molecules (and orbiting electrons) are not moving and have no volume.

These rules represent ideal gases, providing only an approximate representation of their actual behavior.

However, their description is remarkably accurate.

From these assumptions, laws are derived to describe the behavior of gases.

Boltzmann's constant

The final assumption (5) for KE is expressed by:

$$KE = \tfrac{1}{2}\,mv^2$$

$$\tfrac{1}{2}\,mv^2 = (3/2)\,k_{\mathrm{B}}T$$

so

$$KE = (3/2)\,k_{\mathrm{B}}T$$

where KE is kinetic energy, m is mass, v is velocity, T is the temperature (Kelvin), and k_B is *Boltzmann's constant* ($k_{\mathrm{B}} = 1.38 \times 10^{-23}$ J/K).

Boltzmann's constant relates macroscopic and microscopic behavior of gases to their particles.

The $KE = (3/2)\,k_{\mathrm{B}}T$ equation is significant because it states that a gas particle's average kinetic energy is proportional to its absolute temperature.

Increasing the temperature of the gas particles increases the total speed and overall energy.

Gas atoms are infinitesimally small; nearly impossible to accurately measure a particle's speed.

The speed of gas particles is defined by the root-mean-square speed (u_{rms}).

Root-mean-square speed equation:

$$u_{\mathrm{rms}} = \sqrt{\frac{3RT}{MM}}$$

where R is the ideal gas constant, T is the absolute temperature (K), and MM is the molar mass of the gas sample.

Root-mean-square speed written with the *Boltzmann constant*, k_{B}:

$$u_{\mathrm{rms}} = \sqrt{\frac{3kT}{m}}$$

where k is the Boltzmann constant, T is the absolute temperature (K), and m is the mass of one gas molecule.

Boltzmann distribution plot

A plot of the distribution of speeds for each gas particle at a given temperature shows a slightly asymmetric curve.

This speed-distribution curve is the *Maxwell-Boltzmann distribution curve.*

Boltzmann distribution plot shows that the curve's peak corresponds to the most probable speed.

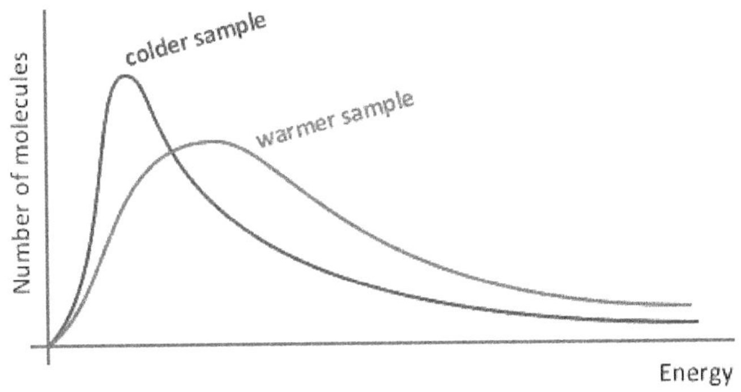

Maxwell-Boltzmann Distribution curves of molecular speeds at two temperatures

If the temperature (K) remains constant, the total kinetic energy remains unchanged.

However, the energy is distributed in several ways, and the gas particles travel at many speeds.

This distribution changes continually as the gas atoms collide with each other and with the container walls.

The curve flattens (i.e., net area under the curve is greater) and shifts to the right at higher temperatures, indicating that more gas molecules move at higher speeds and therefore possess more kinetic energy.

Effusion and diffusion

Effusion is the flow of gas particles *under pressure* from one compartment through a small opening, as shown.

Diffusion is the process whereby a substance (solute or particle) *spreads* from a region of high concentration to one of lower concentration. Diffusion is explained using the kinetic theory of motion.

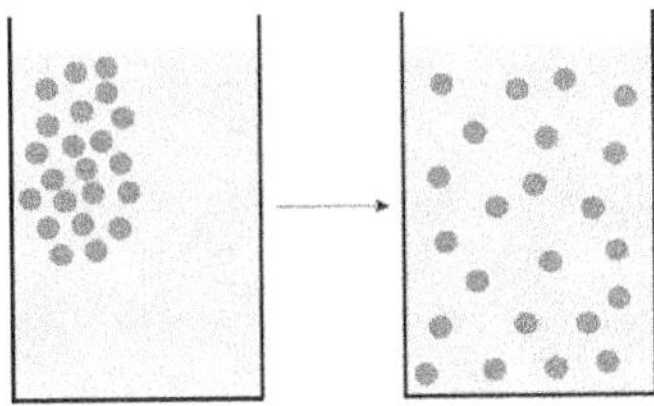

Diffusion of solutes in a solvent

According to the kinetic theory, heavier gases diffuse more slowly than lighter gases due to the difference in their molecular speeds.

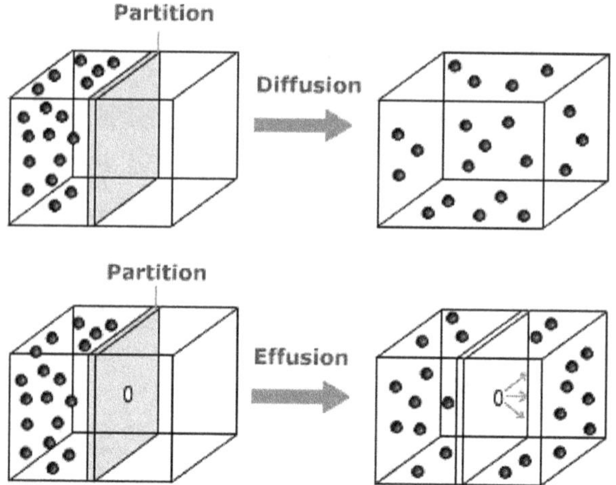

Diffusion and effusion of gas particles

Graham's Law

Scottish chemist Thomas Graham (1805-1869) developed *Graham's Law*, which states that *at constant temperature and pressure conditions, the rates at which two gases diffuse are inversely proportional to the square root of their molar masses.*

Graham's Law is expressed by:

$$\frac{r_1}{r_2} = \sqrt{\frac{MM_2}{MM_1}}$$

where r_1 and r_2 are the diffusion rates of gas 1 and gas 2, respectively, and MM_1 and MM_2 are the molar masses of the gases.

Graham's Law is a direct outcome of the KMT as it explains why gases diffuse and effuse at different rates.

Graham's Law applies to the *effusion* of gas particles. The equation for *effusion* is the same as for *diffusion.*

Gas Laws

Ideal gas law

The *ideal gas law* was proposed in 1834 by French physicist Émile Clapeyron (1799-1864). It is a combination of the gas laws discussed, and it is the equation regarding the state of hypothetical gas.

Ideal gas law explains relationship between pressure (P), volume (V), and temperature (T).

The gas law is expressed as:

$$PV = nRT$$

where P is the pressure, V is the volume, T is the temperature, n is the number of moles, and R is the gas constant (use $R = 0.082$ L·atm/K·mol).

One mole $= 6.023 \times 10^{23}$ molecules; technically, the number of H atoms in one gram of H.

Because atoms are so small, it is more practical to count atoms in moles than to count them individually.

The ideal gas law applies to the pressure exerted by a gas on a cylinder with a moving wall.

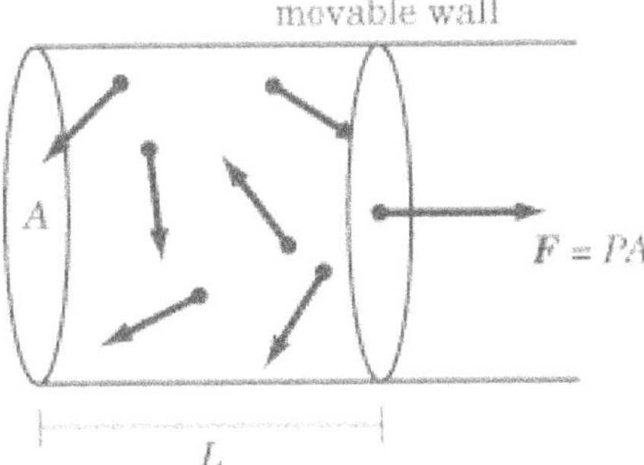

Since pressure is equal to $P = F/A$, the force that the gas exerts on the wall is equal to $F = PA$.

If this force moves the wall back a length of L, volume of the cylinder increases by $\Delta V = LA$.

By solving for A and substituting it into the equation for force, the result is $F = P\Delta V L$, equal to $P\Delta V = FL$.

For example, suppose 2.5 moles of gas are at standard temperature and pressure. What is the volume occupied by the gas? (Use standard pressure $= 1.00$ atm and standard temperature $= 273.0$ K).

Use the ideal gas law:

$$PV = nRT$$

Solve for V:

$$V = (nRT) / P$$

$$V = [(2.5 \text{ mol}) \cdot (0.082 \text{ L} \cdot \text{atm/K} \cdot \text{mol}) \cdot (273.0 \text{ K})] / (1.00 \text{ atm})$$

$$V = 56 \text{ L}$$

Several useful expressions are derived from the ideal gas law (PV = nRT), which relates the molar mass and density of gases to pressure and temperature, by substituting a known expression for one of the ideal gas law variables.

Work (w) is *force multiplied by the distance traveled*:

$$w = Fd$$

By pushing the wall a distance of L with a force of F, the gas has done work equal to FL.

When gas does work, it symbolizes a change in energy.

If a change in PV is equal to a change in energy, then PV is the total energy of the gas.

For the ideal gas law, this means that nRT is the expression for the total kinetic energy of the gas molecules.

The ideal gas law (PV = nRT) is related to the number of molecules (N) and the Boltzmann's constant (k), which has a value of 1.381×10^{-23} J/K:

$$PV = NkT$$

The number of moles (n) and the gas constant (R) are constant.

Other gas laws derive from the ideal gas law by holding specific variables constant:

 Boyle's Law

 Charles' Law

 Combined Gas Law

 Closed Container Law

Ideal gases follow KMT (isothermal process)

Kinetic Molecular Theory (KMT) described previously is based on the theoretical concept of an ideal gas.

Ideal gases follow the _five_ assumptions of KMT, whereby gas molecules:

> move in _random motion_

> are _point masses_ with _no volume_

> have _no intermolecular forces_

> exhibit _elastic collisions_

> kinetic energy (_KE_) of molecules is _proportional to temperature_

Compared to an ideal gas, the behavior of a real gas is _complicated._

By conceptualizing ideal gas behavior, real gas behavior is easier to comprehend.

In the 17th and 18th centuries, scientists discovered several relationships between the properties of gases.

Boyle's, Charles', and Avogadro's Laws relate the properties of pressure, volume, and temperature, which were developed empirically as individual cases of the ideal gas equation: $PV = nRT$.

Boyle's Law (isothermal process)

In 1661, Robert Boyle (1627-1691) studied the compressibility of gases. He observed that _the volume of a fixed amount of gas at a given temperature is inversely proportional to the pressure exerted on the gas._

For _Boyle's Law_, the temperature of a gas is held constant.

It states that an _increase in pressure_ causes a _decrease in volume._

Likewise, a _decrease in pressure_ causes an _increase in volume._

Boyle's Law is expressed as:

$$P_1 V_1 = P_2 V_2$$

where subscripts 1 and 2 refer to the same gas sample under two sets of pressure and volume conditions.

A plot of volume _vs._ pressure for a gas illustrates that Boyle's Law is a particular case of the ideal gas law, where _n_ (moles of gas) and _T_ (temperature) are constant (figure below).

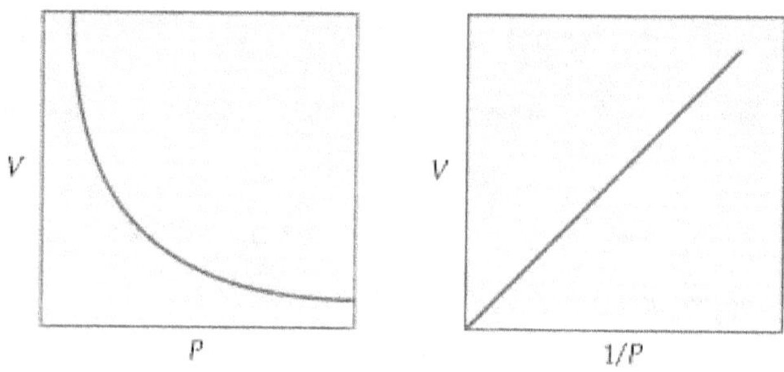

Boyle's Law: as pressure increases, volume decreases

For example, what is the new pressure in a container of gas in a 15.0 L container at a pressure of 5.00 atm when the volume decreases to 0.500 L?

Substitute the known quantities (P_1, V_1, and V_2) into Boyle's Law equation to solve for P_2.

Note: Volume units must be consistent. For example, volumes are expressed in units of liters.

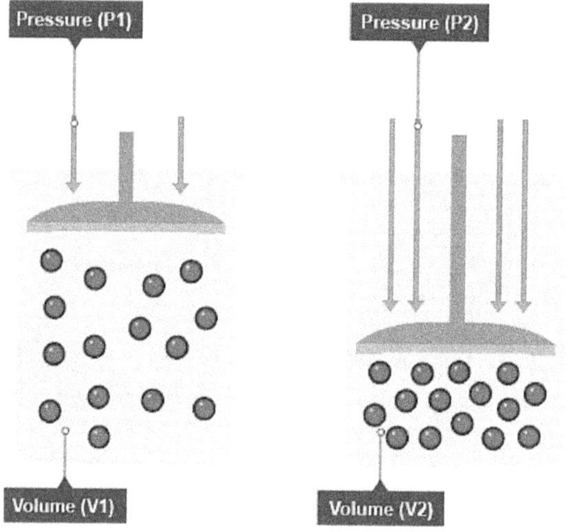

Boyle's Law:

$$P_1V_1 = P_2V_2$$

Solve for P_2:

$$P_1V_1 / V_2 = P_2$$

$$[(5.00 \text{ atm}) \cdot (15.0 \text{ L})] / 0.500 \text{ L} = P_2$$

$$P_2 = 150 \text{ atm}$$

Charles' Law (isobaric process)

French scientist Jacques Charles (1746-1823) developed the relationship between temperature and gas volume.

Charles' Law states that *at constant pressure, the volume of a gas is directly proportional to its absolute temperature (Kelvin)*. It describes the behavior of a gas under *constant pressure*.

Volume and temperature are directly proportional; that is, when the temperature increases, the volume increases, and when the temperature decreases, the volume decreases.

Charles' Law states that the volume of a gas is proportional to temperature

Charles' Law is expressed as:

$$\frac{V_1}{T_1} = \frac{V_2}{T_2}$$

where subscripts 1 and 2 refer to the same gas sample but under different temperature and volume conditions.

Temperature vs. volume plot for a gas illustrates that Charles' Law is another application of the ideal gas law, where n and P are constant, as shown.

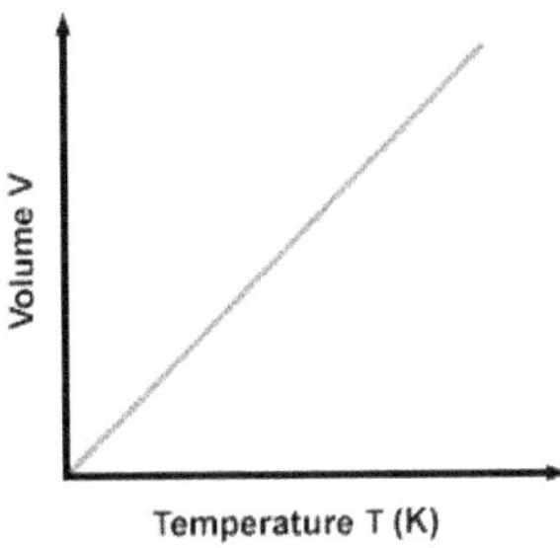

Charles' Law states that as temperature increases, volume increases

For example, if there is 25.0 L of gas at 0 °C, and the temperature is raised to 100 °C, what is the volume of the gas?

First, the temperature is converted to Kelvin:

$$T_1 = 0 \text{ °C} + 273 = 273 \text{ K}$$

$$T_2 = 100 \text{ °C} + 273 = 373 \text{ K}$$

Substitute the known quantities into the equation for Charles' Law:

$$V_1 / T_1 = V_2 / T_2$$

Rearrange and solve for V_2:

$$V_1 T_2 / T_1 = V_2$$

$$(25.0 \text{ L}) \cdot (373 \text{ K}) / 273 \text{ K} = V_2$$

$$V_2 = 34.2 \text{ L}$$

Gay-Lussac's Law

In 1802, French chemist Louis Gay-Lussac (1778-1850) proposed *Gay-Lussac's Law*, which states that the *pressure* of a given sample of gas is directly proportional to its *absolute temperature* (Kelvin) at a *constant volume*.

Gay-Lussac's Law is expressed by:

$$\frac{P_1}{T_1} = \frac{P_2}{T_2}$$

where subscripts 1 and 2 refer to the same gas sample but under different temperature and pressure conditions.

A plot of *temperature vs. pressure* for a gas illustrates that Gay-Lussac's Law is another particular case of the ideal gas law, where n and V are constant.

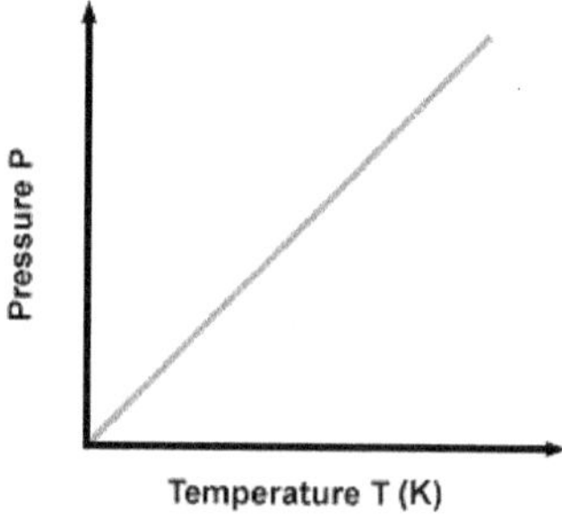

Gay-Lussac's Law states that as temperature increases, pressure increases

For example, suppose a gas is at 30.0 atm pressure and 100 °C, and the temperature changes to 400 °C. What is the new pressure of the gas?

Convert temperature to Kelvin units.

$$T_1 = 100\ °C + 273 = 373\ K$$

$$T_2 = 400\ °C + 273 = 673\ K$$

Gay-Lussac's Law solves P_2 by substituting the quantities into the equation.

$$\frac{P_1}{T_1} = \frac{P_2}{T_2}$$

$$P_2 = [(30.0\ \text{atm}){\cdot}(673\ K)] / 373\ K$$

$$P_2 = 54.1\ \text{atm}$$

Closed container law (constant volume)

The *closed container law* describes the behavior of a gas under constant volume. In such cases, pressure and temperature are directly proportional:

$$\frac{P_i}{T_i} = \frac{P_f}{T_f}$$

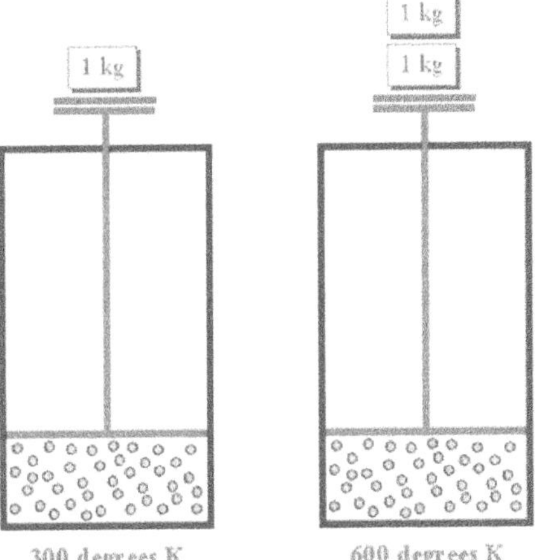

Avogadro's Law

Volume of gas in a container is affected by pressure, temperature, and number of moles of gas.

The relationship between the quantity of gas and its volume was derived by Italian scientist Amadeo Avogadro (1776-1856).

Avogadro's Law states that *gases at a given temperature and pressure occupying a volume are directly proportional to the number of moles of gas present.*

Avogadro's Law is expressed as:

$$\frac{n_1}{V_1} = \frac{n_2}{V_2}$$

where n_1 and n_2 are moles of gas 1 and gas 2, and V_1 and V_2 are the volumes of gas 1 and gas 2.

Volume vs. moles plot of gas (shown below) illustrates that Avogadro's Law is a particular case of the ideal gas law where T and P are held constant.

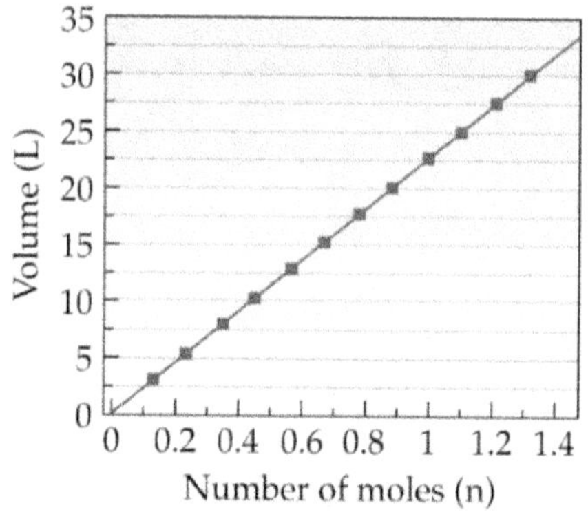

Avogadro's Law states that as moles of a gas increase, volume increases

For example, if 8.00 moles of a gas occupy a volume of 4.00 L at constant pressure and temperature, what volume of gas would 16.0 moles of this gas occupy at the same temperature and pressure?

Use *Avogadro's Law*

$$n_1 / V_1 = n_2 / V_2$$

where n_1 and n_2 are the moles of gas 1 and 2, and V_1 and V_2 are the volumes of gas 1 and gas 2.

Solve for the V_2:

$$V_2 = (V_1 \times n_2) / n_1$$

$$V_2 = [(4.00 \text{ L}) \cdot (16.0 \text{ mol})] / 8.00 \text{ mol}$$

$$V_2 = 8 \text{ L}$$

Combined gas law

Combined gas law is an amalgam of the other gas laws.

The combined gas law is *not* the ideal gas law (PV = nRT); it is an equation that relates changes in temperature, volume, and gas pressure.

Combined gas law is:

$$\frac{P_1 V_1}{T_1} = \frac{P_2 V_2}{T_2}$$

where subscripts 1 and 2 refer to the same gas sample but under different temperatures, pressures, and volumes.

For example, suppose a gas is at 15.0 atm pressure, with a volume of 25.0 L and a temperature of 300.0 K. What would the volume of the gas be at standard temperature and pressure? (Use standard pressure = 1.00 atm and standard temperature = 273.0 K)

Use the *combined gas law* at standard temperature (T_2) and pressure (P_2):

$$(P_1 V_1) / T_1 = (P_2 V_2) / T_2$$

Solve for V_2:

$$(P_1 \times V_1 \times T_2) / (T_1 \times P_2) = V_2$$

$$V_2 = [(15.0 \text{ atm}){\cdot}(25.0 \text{ L}){\cdot}(273.0 \text{ k})] / [(300.0 \text{ K}){\cdot}(1.00 \text{ atm})]$$

$$V_2 = 341 \text{ L}$$

Molar mass of a gas

Moles of a gas *n* can be expressed as the *mass of gas* in grams over the *molar mass* of gas:

$$n = m / MM$$

where *MM* = molar mass, *m* = mass of the gas in grams, and *n* = moles of gas.

Substitute expression (*m/MM*) into the ideal gas law for moles (*n*):

$$PV = nRT$$

$$PV = mRT / MM$$

Multiplying each side by the molar mass (*MM*):

$$(MM)PV = mRT$$

This derived equation determines the molar mass of a gas from experimental data, where the gas' mass, pressure, volume, and temperature are measured.

Divide each side of the above expression by the volume (V):

$$(MM)P = gRT \,/\, V$$

Density:

$$\rho = g/V$$

Substitute density for g/V in the derived equation:

$$(MM)P = \rho RT$$

This derived equation relates the gas's pressure, density, and temperature, like the other empirical gas laws.

For example, what is the molar mass of a gas with the density (ρ) of 1.855 g/L at 0.950 atm and 297.0 K.? (Use $R = 0.0820$ L·atm/K·mol)

Substitute the quantities into the derived equation above.

$$(MM)P = \rho RT$$

$$(MM)\cdot(0.950 \text{ atm}) = (1.855 \text{ g/L})\cdot(0.0820 \text{ L·atm/K·mol})\cdot(297.0 \text{ K})$$

Solve for molar mass (MM) by dividing each side of the equation by P (0.95 atm).

$$MM = [(1.855 \text{ g/L})\cdot(0.0820 \text{ L·atm/K·mol})\cdot(297.0 \text{ K})] \,/\, (0.950 \text{ atm})$$

$$MM = 47.6 \text{ g/mol}$$

Deviation of real-gas behavior

Kinetic Molecular Theory and the *ideal gas law* are approximations of gas behavior.

These theories are based on the theoretical ideal gas, while real gases have deviations from this behavior (the five critical assumptions of an ideal gas).

When molecules are far apart (i.e., low pressure and high temperature), real gas behaves more like an ideal gas.

When molecules are brought close (under high pressure and low temperature), gas molecules experience an intermolecular attraction. Therefore, the gas may deviate significantly from the behavior of an ideal gas.

The most substantial deviations from the ideal occur at high pressure and low temperature. The gas molecules are "squished" under these conditions and thus experience intermolecular interactions. Molecular volume becomes significant when the total volume is *compressed*.

Intermolecular attractions cause collisions to be sticky and inelastic.

Furthermore, gases *condense* into liquids at extremely high pressures and low temperatures.

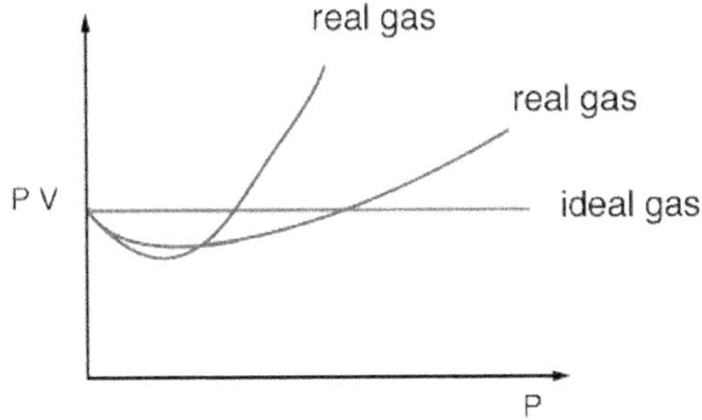

An ideal gas is a *point mass*, a particle so small that its *volume is negligible*.

A real gas particle has a volume because real gases liquefy as they cool and cannot compress to zero volume; the molecules of a gas are not dimensionless points.

Collisions between gas particles are "elastic" with *no attractive or repulsive forces* for an ideal gas.

Thus, no energy is exchanged during the collisions.

For a real gas, collisions are *inelastic* (i.e., objects stick and lose kinetic energy).

Van der Waals equation describes deviations from ideal gas

In 1873, Dutch theoretical physicist Johannes Diderik van der Waals (1837-1923) derived a mathematical relationship that describes the behavior and condensation of real gases in*to the liquid phase*.

Ideal gas equation for comparison:

$$PV = nRT$$

"Corrections" are made to the pressure (P) and volume (V) terms,

where P becomes:

$$P + (n^2a) / V^2$$

and V becomes:

$$V - nb$$

Van der Waals equation is expressed by:

$$[P + (n^2a) / V^2] \times (V - nb) = nRT$$

Since the collisions of real gases are inelastic, the term n^2a / V^2 corrects the interactions between gas particles.

Since real gas particles occupy space (volume), the nb term corrects the volume neglected by the ideal gas law.

Values of a and b are constant and are determined *experimentally* for each gas. The constants a and b would be given in a problem, and pressure or temperature can be easily solved.

Solving the volume is nontrivial and involves solving a cubic polynomial equation.

For pressure, van der Waals equation is rearranged by dividing by the volume term $(V - nb)$:

$$P + (n^2a) / (V^2) = (nRT) / (V - nb)$$

Subtract the intermolecular interaction term (n^2a / V^2) from both sides of the equation:

$$P = [(nRT) / (V - nb)] - [(n^2a) / (V^2)]$$

Repulsion   Attraction

For example, use the real gas law to find the pressure of 2.00 moles of carbon dioxide (CO_2) gas at 298 K in a 5.00 L container. (Use van der Waals constants for CO_2 whereby $a = 3.592$ $L^2 \cdot atm/mol^2$ and $b = 0.04267$ L/mol).

Substitute the variables into the equation and solve for the pressure (P):

$$P = [(nRT) / (V - nb)] - [(n^2a) / (V^2)]$$

$$P = [(2 \text{ mol}) \cdot (0.0821 \text{ L} \cdot \text{atm}) \cdot (298 \text{ K})] / [(5 \text{ L}) - (2.00 \text{ mol}) \text{mol} \cdot \text{K} \cdot (0.04267 \text{ L})]$$
$$- [(2 \text{ mol})^2 \cdot (3.592 \text{ L}^2 \cdot \text{atm}) / (5.00 \text{ L})^2 \cdot \text{mol}^2]$$

$$P = 9.38 \text{ atm}$$

Compare this value (9.38 atm) to the calculated pressure using the ideal gas law:

$$PV = nRT$$

$$P = nRT / V$$

$$P = [(2.00 \text{ mol}) \cdot (0.0821 \text{ L} \cdot \text{atm}) \cdot (298 \text{ K})] / (5.00 \text{ L}) \text{mol} \cdot \text{K}$$

$$P = 9.77 \text{ atm}$$

$P = 9.38$ atm (van der Waals) does not equal $P = 9.77$ atm (ideal gas)

Although the results from the two equations are similar, they are not identical, illustrating the difference in the behavior of a real gas (one that does not obey the five stated assumptions) compared to an ideal gas.

Partial Pressure

Mole fraction

There is a significant distance between gas molecules, allowing other gas molecules to occupy the same space.

Different gases combine to form homogeneous mixtures.

Each gas behaves *independently* and makes its unique contribution to the total pressure.

Gas's partial pressure is as if it were the *only gas* in the container.

Mole fraction (X_A) indicates the ratio between moles of a specific gas component and the total number of moles in the homogeneous mixture.

The symbol X_A expresses the component gas's mole fraction in the mixture.

$$X_A = \text{(moles of gas A) / (total moles of the gas mixture)}$$

Dalton's Law of partial pressures

When two or more nonreactive gases are present in the same container, they behave relatively independently of each other.

Dalton's Law of partial pressures states that *the total pressure of a mixture of gases equals the sum of the individual gas pressures.*

It states that the total pressure of a mixture of gases equals the sum of the partial pressures of its gas components.

Dalton's Law of partial pressures is expressed by:

$$P_T = P_A + P_B + P_C \ldots$$

where P_T is the total pressure in the container, and P_A, P_B and P_C equal the partial pressures of gases A, B, and C.

The partial pressure of a gas is related to its mole fraction by:

$$P_A = X_A P_T$$

where $X_A = \text{(moles of gas A) / (total moles of gas)}$.

For example, suppose there is 1.0 L of oxygen at 1.0 atm pressure in a container, 1.0 L of nitrogen at 0.5 atm pressure in another container, and 1.0 L of hydrogen at 3.0 atm pressure in a third container. What is the total pressure if the gases combine in a single 1.0 L container?

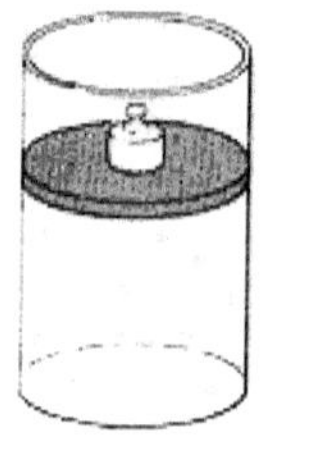 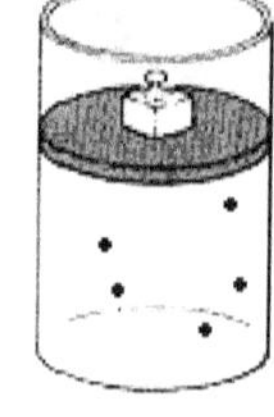 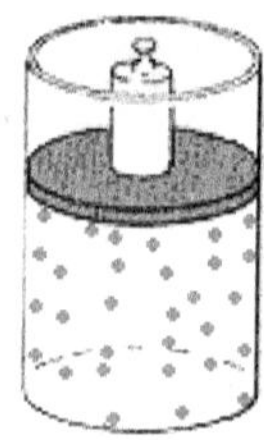

$P_{O_2} = 1.0$ atm $P_{N_2} = 0.5$ atm $P_{H_2} = 3.0$ atm

For a single 1.0 L container, the total pressure is the sum of the partial pressures (X_X) of each gas component:

$X_X = 1.0$ atm $+ 0.5$ atm $+ 3.0$ atm $= 4.5$ atm

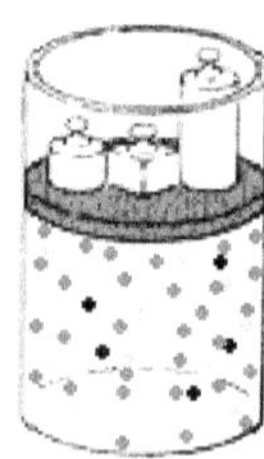

$P_{Total} = P_{O_2} + P_{N_2} + P_{H_2}$

$P_{Total} = 4.5$ atm

Measuring vapor pressure

A critical application of Dalton's Law of partial pressure is to determine *vapor pressure* (i.e., the pressure exerted by a vapor in thermodynamic equilibrium with its liquid or solid phase).

Vapor pressure determines the amount of a water-insoluble gaseous reaction product or a slightly soluble gas, such as hydrogen or oxygen.

A typical experimental method (below) for determining the amount of gas present involves collecting it over water and measuring the height of the displaced water. This is accomplished by placing a tube into an inverted bottle, the opening of which is immersed in a larger container of water. The gas bubbles into the test tube, displacing the water until the test tube is full.

The collected gas is not the only gas in the test tube, as liquid water is in equilibrium with its vapor. Therefore, the collected gas is a mixture of two gases: the gas collected and water vapor.

Partial pressure of water is the *vapor pressure of water* and is dependent on temperature.

Volume of gas collected consists of a mixture of the *gas collected* and *water vapor*.

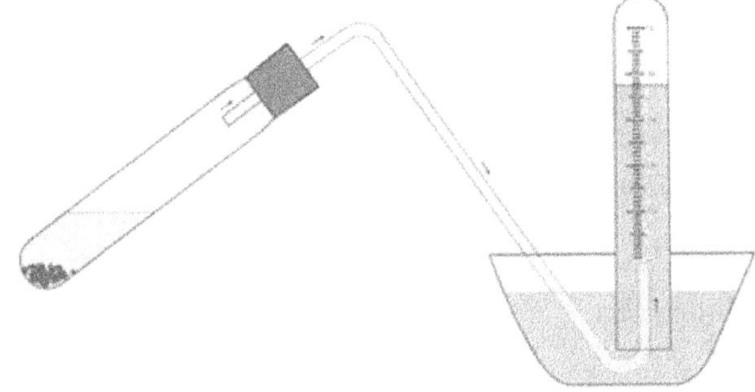

An experimental method to determine the amount of gas present

Total pressure is the sum of the two contributing partial pressures.

Pressure of dry gas collected is calculated from a table of water vapor pressure values at various temperatures. The partial pressure of the dry gas is calculated by subtracting the reference vapor pressure of H_2O.

The equalizing of the atmospheric pressure yields the partial pressure of the gaseous product collected. The amount of gas products can be determined by the known volume and temperature.

Vapor pressure of water at specific temperatures

Temperature (°C)	Vapor Pressure (kPa)	Temperature (°C)	Vapor Pressure (kPa)
0	0.61	26	3.36
5	0.87	27	3.57
10	1.23	28	3.78
15	1.71	29	4.00
16	1.82	30	4.24
17	1.94	35	5.62
18	2.06	40	7.38
19	2.20	45	9.58
20	2.34	50	12.33
21	2.49	60	19.92
22	2.64	70	31.16
23	2.81	80	47.34
24	2.98	90	70.10
25	3.17	100	101.30

Determining partial pressure

For example, a small piece of zinc (Zn) reacts with dilute hydrochloric acid (HCl) to form hydrogen (H_2) gas, which is collected over water at 16.0 °C. The total pressure is adjusted to a barometric pressure of 100.24 kPa, and the volume of hydrogen gas is measured as 1,495 cm^3. What are the partial pressure and mass of hydrogen gas? (Use $R = 8.314$ L·kPa·K^{-1}·mol^{-1})

Partial pressure of hydrogen gas:

$$P_{HCl} = 100.24 \text{ kPa} - 1.82 \text{ kPa}$$

$$P_{HCl} = 98.42 \text{ kPa}$$

To calculate the mass of the hydrogen gas, use the ideal gas equation (PV = nRT) and solve the number of moles.

For hydrogen (H_2) gas, multiply the mass of one hydrogen atom (1.008 g/mol) by two to obtain the molecular mass of 2.016 g/mol.

$$H_2 = 2 \times 1.008 \text{ g/mol}$$

$$H_2 = 2.016 \text{ g/mol}$$

Temperature and volume were given.

The partial pressure of hydrogen has been previously calculated; therefore, the ideal gas equation can be applied.

Determine the number of moles of H_2 gas:

$$PV = nRT$$

$$n = PV / RT$$

$$n = [(98.42 \text{ kPa}) \cdot (1.495 \text{ L})] / [(8.314 \text{ L·kPa·}K^{-1}\text{·}mol^{-1}) \cdot (289 \text{ K})]$$

$$n = 0.061 \text{ mol of } H_2 \text{ gas}$$

Calculate the mass of the H_2 gas formed:

$$\text{mass} = (\text{moles of } H_2 \text{ gas}) \times (\text{molecular mass of } H_2 \text{ gas})$$

$$\text{mass} = 0.061 \text{ mol } H_2 \text{ gas} \times 2.016 \text{ g/mol } H_2 \text{ gas}$$

$$\text{mass} = 0.123 \text{ g of } H_2 \text{ gas}$$

Intermolecular Forces

Forces between molecules

Intermolecular forces are *attractive or repulsive* forces between molecules (e.g., hydrogen bonding, London dispersion).

However, some other intermolecular interactions are not as strong as covalent or ionic bonds, but they still play a significant role in determining substances' properties.

Dipole interactions

Molecules with dipole moments (from differences in electronegativity of bonded atoms) are attracted to each other. The more polar the molecule is, the stronger the dipole attraction.

Four types of dipole attractions:

- *Ion-dipole* attractions form between an ion and a polar molecule. Since a polar molecule has a slight charge (i.e., dipole) and an ion is a charged atom, these particles are attracted to each other.

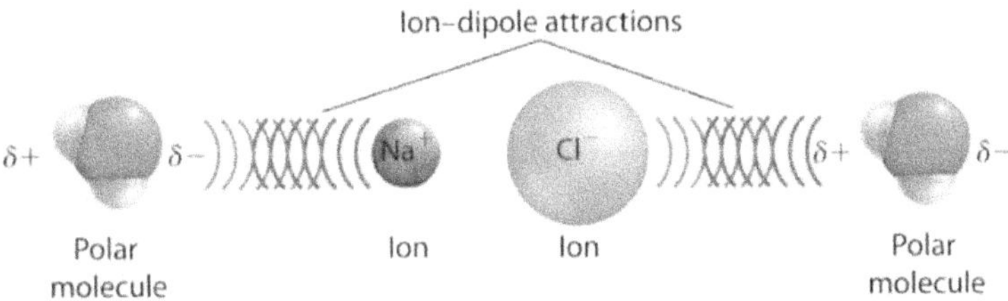

- *Dipole-dipole* attraction forms between two polar molecules. The partially positive side of one polar molecule attracts the partially negative side of another polar molecule.

- *Dipole-induced dipole* attraction forms between partially polar and nonpolar molecules.

  For example, the polar H_2O molecules induce the nonpolar O_2 molecules, shifting electron density and inducing temporary partial positive (δ^+) and negative (δ^-) regions in the nonpolar O_2 molecules.

  positive pole ⊢────► negative pole

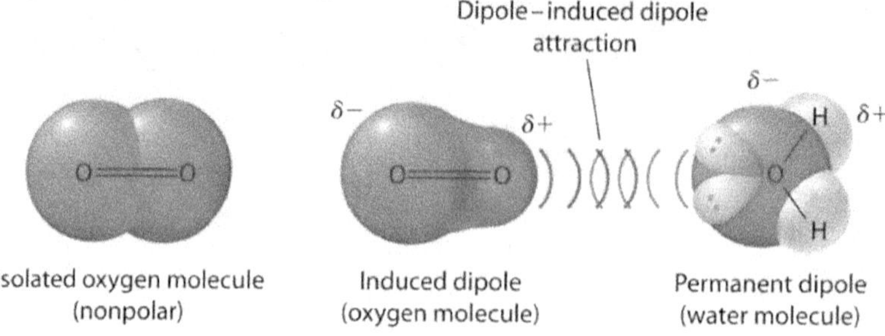

Isolated oxygen molecule
(nonpolar)

Induced dipole
(oxygen molecule)

Permanent dipole
(water molecule)

- ***Induced dipole-induced dipole*** *attraction* (or *London dispersion force*) forms between two (or more) nonpolar molecules. These nonpolar molecules become temporarily polar (i.e., momentary flux) due to random, instantaneous induction by another polar molecule that results in the probability of finding the negative electron in a specific region within its orbital.

Longer molecules have a greater surface area and experience higher London dispersion forces.

For example, octane (C_8H_{18}) has a stronger London dispersion force than methane (CH_4).

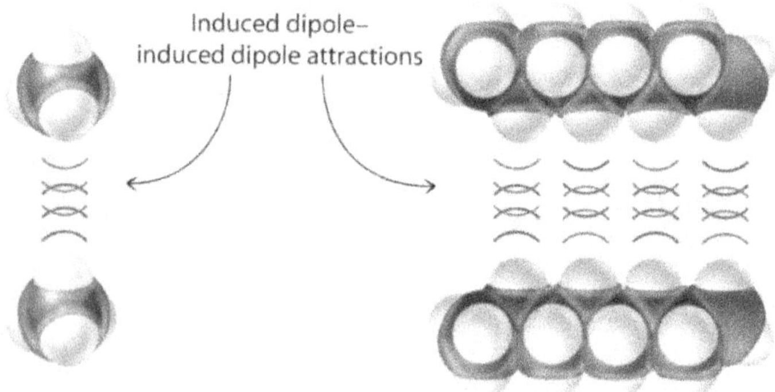

Molecular attractions involving dipoles

Attraction	Relative Strength
Ion–dipole	Strongest
Dipole–dipole	
Dipole–induced dipole	
Induced dipole–induced dipole	Weakest

Hydrogen bonding

Hydrogen bonding is a specific type of bond that, based on its strength, is categorized as a *dipole-dipole* attraction.

Hydrogen (hydrogen bond donor with one valence electron) bonding occurs when hydrogen is covalently bonded directly to the electronegative atom of N, O, or F (hydrogen bond donors).

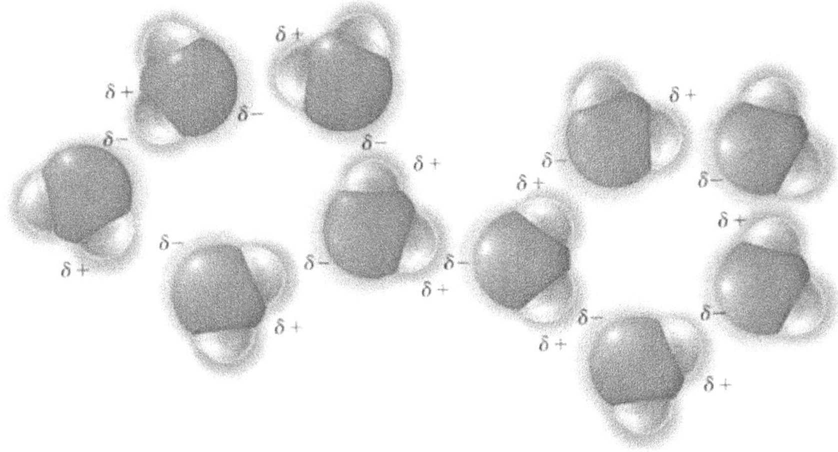

Hydrogen bonding produces differences in physical properties

Bonding electrons are closer to the more electronegative N, O, or F atom.

Due to these factors, the hydrogen nucleus is exposed.

Highly positive hydrogen nucleus attracts the *electronegative atom* from a neighboring molecule. The neighboring molecule needs a lone electron pair to form a hydrogen bond. This creates a powerful intermolecular force.

The more polar a bond is, the stronger the hydrogen bond.

H–F bonds are the most polar, then H–O bonds, and then H–N bonds.

Intermolecular forces in polar molecules

Hydrogen bonding significantly increases the boiling point of compounds.

For example, methanol (CH_3OH) and methane (CH_4) have similar molecular structures.

However, methanol is liquid at room temperature, while methane is a gas.

Methanol is liquid at room temperature because of hydrogen bonding between methanol molecules, which is *much stronger* than the London dispersion (weak intermolecular) forces between the nonpolar methane molecules.

Therefore, methane is a gas at room temperature.

Hydrogen bond donors and acceptors

The molecule bonding (by lone pair electrons) to the hydrogen (bonded to F, O, N) is the hydrogen bond *acceptor*.

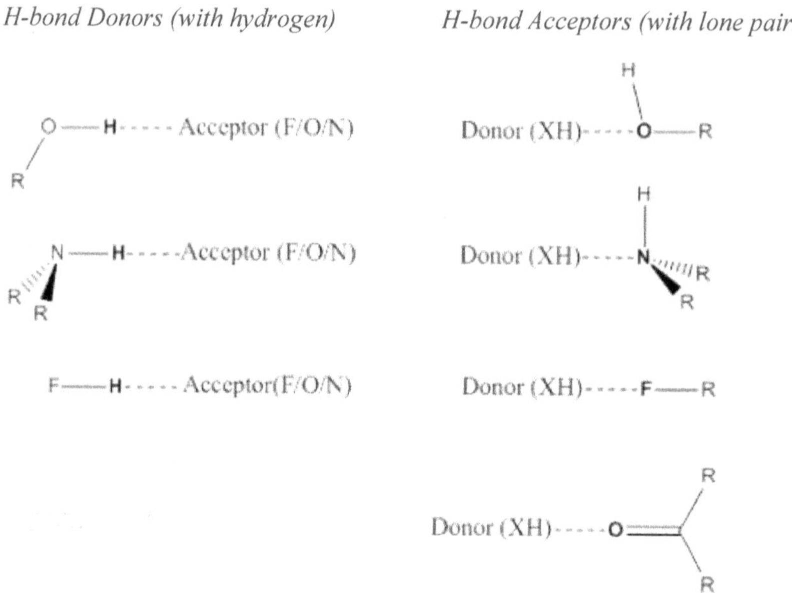

Hydrogen bond *donor* has a δ^+ hydrogen bonded to an electronegative F, O, or N.

Hydrogen bond *acceptor* has a *lone pair of electrons* to bond to the δ^+ hydrogen of the donor.

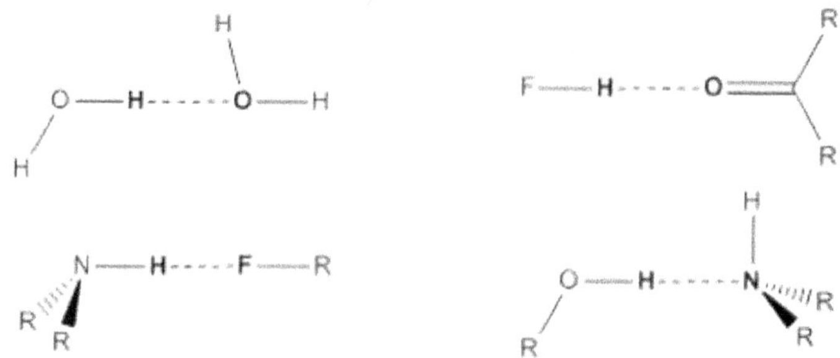

Examples of hydrogen-bonded molecules

Van der Waals' forces

Atoms and molecules exhibit weak intermolecular attractions, known as *van der Waals forces*. This results from a molecule's nucleus having a weak attraction to another molecule's valence electrons.

Van der Waals equation (above) accounts for these *attractive intermolecular forces*.

Van der Waals force affects the *boiling point* because the boiling point reflects the kinetic energy required to break a molecule's cooperative attractions within the liquid.

Larger *molecular mass* reduces the kinetic energy ($KE = \frac{1}{2}mv^2$) for the same energy input.

London dispersion forces

Nonpolar molecules have *fluctuating dipoles* that align with other molecules from one instant to the next.

Momentary flux of electron density is attracted to nearby nuclei's positive charge, and this attraction decreases dramatically as the distance between the molecules increases.

London dispersion forces are a subset of van der Waals' attractive forces resulting from temporarily fluctuating molecules ("induced dipole or induced dipole attraction").

London dispersion forces exist for many molecules but are a significant source of intermolecular forces for nonpolar molecules that lack dipole-dipole, dipole-induced dipole, or hydrogen bonding interactions.

London dispersion forces are weak compared to the forces generated between polar molecules.

For polar molecules, dipole forces are strong, influential, and predominant.

Generally, larger nonpolar molecules have higher boiling points than smaller nonpolar molecules due to the increased London dispersion forces.

Branched molecules (i.e., hydrocarbons) have weaker London dispersion forces.

Notes for active learning

Notes for active learning

Notes for active learning

Notes for active learning

PRACTICE QUESTIONS
&
DETAILED EXPLANATIONS

Practice Set 1: Questions 1–20

1. Which statement is NOT true regarding vapor pressure?

> I. Solids do not have a vapor pressure
>
> II. Vapor pressure of a pure liquid does not depend on the amount of vapor present
>
> III. Vapor pressure of a pure liquid does not depend on the amount of liquid present

A. I only

B. II only

C. III only

D. I and II only

E. I, II and III

2. How does the volume of a fixed sample of gas change if the temperature is doubled at constant pressure?

A. Decreases by a factor of 2

B. Increases by a factor of 4

C. Doubles

D. Remains the same

E. Requires more information

3. When a solute is added to a pure solvent, the boiling point [   ] and freezing point [   ]?

A. decreases … decreases

B. decreases … increases

C. increases … decreases

D. increases … increases

E. remains the same … remains the same

4. Consider the phase diagram for H_2O. The termination of the gas-liquid transition at which distinct gas or liquid phases do NOT exist is the:

A. critical point

B. endpoint

C. triple point

D. condensation point

E. inflection point

5. The van der Waals equation of state for a real gas is expressed as $[P + n^2a / V^2]\cdot(V - nb) = nRT$. The van der Waals constant, a, represents a correction for:

A. negative deviation in the measured value of P from that of an ideal gas due to the attractive forces between the molecules of a real gas

B. positive deviation in the measured value of P from that of an ideal gas due to the attractive forces between the molecules of a real gas

C. negative deviation in the measured value of P from that of an ideal gas due to the finite volume of space occupied by molecules of a real gas

D. positive deviation in the measured value of P from that of an ideal gas due to the finite volume of space occupied by molecules of a real gas

E. positive deviation in the measured value of P from that of an ideal gas due to the finite mass of the molecules of a real gas

6. What is the value of the ideal gas constant, expressed in units (torr $\times$ mL) / mole $\times$ K?

A. 62.4

B. 62,400

C. 0.0821

D. 1 / 0.0821

E. 8.21

7. If both the pressure and the temperature of a gas are halved, the volume is:

A. halved

B. the same

C. doubled

D. quadrupled

E. decreased by a factor of 4

8. If container X is occupied by 2.0 moles of O_2 gas, while container Y is occupied by 10.0 grams of N_2 gas, and both containers are maintained at 5.0 °C and 760 torrs, then:

A. container X must have a volume of 22.4 L

B. the average kinetic energy of molecules in X equals the average kinetic energy of molecules in Y

C. container Y must be larger than container X

D. the average speed of the molecules in container X is greater than that of the molecules in container Y

E. the number of atoms in container Y is greater than the number of atoms in container X

9. Among the following choices, how tall should a properly designed Torricelli mercury barometer be?

A. 100 in

B. 380 mm

C. 76 mm

D. 400 mm

E. 800 mm

10. When volatile solvents X and Y are mixed in equal proportions, heat is released to the surroundings. If pure X has a higher boiling point than pure Y, which of the following statements is NOT true?

A. The vapor pressure of the mixture is lower than that of pure Y

B. The vapor pressure of the mixture is lower than that of pure X

C. The boiling point of the mixture is lower than that of pure X

D. The boiling point of the mixture is lower than that of pure Y

E. Not enough information is provided

11. For the balanced reaction $2\,Na + Cl_2 \rightarrow 2\,NaCl$, which of the following is a gas?

    I. Na   II. Cl_2 III. NaCl

A. I only

B. II only

C. III only

D. I and II only

E. I, II and III

12. How does a real gas deviate from an ideal gas?

    I. Molecules occupy a significant amount of space

    II. Intermolecular forces may exist

    III. Pressure is created from molecular collisions with the walls of the container

A. I only

B. II only

C. I and II only

D. II and III only

E. I, II and III

13. A 2.75 L sample of He gas has a pressure of 0.950 atm. What is the pressure of the gas if the volume is reduced to 0.450 L?

A. 5.80 atm

B. 0.520 atm

C. 0.230 atm

D. 0.960 atm

E. 3.40 atm

14. Avogadro's law, in its alternate form, is very similar in form to:

> I. Boyle's law
>
> II. Charles' law
>
> III. Gay-Lussac's law

A. I only

B. II only

C. III only

D. II and III only

E. I, II and III

15. What is the relationship between the pressure and volume of a fixed amount of gas at constant temperature?

A. directly proportional

B. equal

C. inversely proportional

D. decreased by a factor of 2

E. none of the above

16. What is the term for a change of state from a liquid to a gas?

A. vaporization

B. melting

C. deposition

D. condensing

E. sublimation

17. The combined gas law can NOT be written as:

A. $V_2 = V_1 \times P_1 / P_2 \times T_2 / T_1$

B. $P_1 = P_2 \times V_2 / V_1 \times T_1 / T_2$

C. $T_2 = T_1 \times P_1 / P_2 \times V_2 / V_1$

D. $V_1 = V_2 \times P_2 / P_1 \times T_1 / T_2$

E. none of the above

18. Which singular molecule is most likely to show a dipole-dipole interaction?

A. CH_4 **B.** $H{-}C{\equiv}C{-}H$ **C.** SO_2 **D.** CO_2 **E.** CCl_4

19. What happens if the pressure of a gas above a liquid increases, such as by pressing a piston above a liquid?

    A. Pressure goes down, and the gas moves out of the solvent
    B. Pressure goes down, and the gas goes into the solvent
    C. The gas is forced into solution, and the solubility increases
    D. The solution is compressed, and the gas is forced out of the solvent
    E. The amount of gas in the solution remains constant

20. Which of the following two variables are present in mathematical statements of Avogadro's law?

    A. P and V   **C.** n and T
    B. n and P   **D.** V and T
                          E. n and V

Notes for active learning

Practice Set 2: Questions 21–40

21. Which of the following acids has the lowest boiling point elevation?

Monoprotic Acids	Ka
Acid I	1.4×10^{-8}
Acid II	1.6×10^{-9}
Acid III	3.9×10^{-10}
Acid IV	2.1×10^{-8}

A. I

B. II

C. III

D. IV

E. Requires more information

22. An ideal gas differs from a real gas because the molecules of an ideal gas have:

A. no attraction to each other

B. no kinetic energy

C. molecular weight equal to zero

D. appreciable volumes

E. none of the above

23. Which of the following is the definition of standard temperature and pressure?

A. 0 K and 1 atm

B. 298.15 K and 750 mmHg

C. 298.15 K and 1 atm

D. 273 °C and 750 torr

E. 273.15 K and 10^5 Pa

24. At constant volume, as the temperature of a gas sample is decreased, the gas deviates from ideal behavior. Compared to the pressure predicted by the ideal gas law, actual pressure would be:

A. higher, because of the volume of the gas molecules

B. higher, because of intermolecular attractions between gas molecules

C. lower, because of the volume of the gas molecules

D. lower, because of intermolecular attractions among gas molecules

E. higher, because of intramolecular attractions between gas molecules

25. A flask contains a mixture of O_2, N_2 and CO_2. The pressure exerted by N_2 is 320 torr, and CO_2 is 240 torr. If the pressure of the gas mixture is 740 torr, what is the percent pressure of O_2?

    A. 14%   **B.** 18%   **C.** 21%   **D.** 24%   **E.** 29%

26. A balloon contains 40 grams of He with a pressure of 1,000 torrs. When He is released from the balloon, the new pressure is 900 torr, and the volume is half the original. How many grams of He remains in the balloon if the temperature is the same?

    A. 18 grams   **C.** 10 grams

    B. 22 grams   **D.** 40 grams

                                    E. 28 grams

27. Which of the following is a true statement regarding evaporation?

    A. Increasing the surface area of the liquid decreases the rate of evaporation

    B. The temperature of the liquid changes during evaporation

    C. Decreasing the surface area of the liquid increases the rate of evaporation

    D. Molecules with greater kinetic energy escape from the liquid

    E. Not enough information is provided to make any conclusions

28. Under which conditions does a real gas behave most nearly like an ideal gas?

    A. High temperature and high pressure

    B. High temperature and low pressure

    C. Low temperature and low pressure

    D. Low temperature and high pressure

    E. If it remains in the gaseous state regardless of temperature or pressure

29. Which of the following compounds has the highest boiling point?

    A. CH_3OH

    B. $CH_3CH_2CH_2CH_2CH_2OH$

    C. $CH_3OCH_2CH_2CH_2CH_3$

    D. $CH_3CH_2OCH_2CH_2CH_3$

    E. $CH_3CH_2CH_2C(OH)HOH$

30. According to the kinetic theory of gases, which of the following is the average kinetic energy of the gas particles directly proportional to?

A. temperature

B. molar mass

C. volume

D. pressure

E. number of moles of gas

31. Under ideal conditions, which of the following gases is least likely to behave as an ideal gas?

A. CF_4 B. CH_3OH C. N_2 D. O_3 E. CO_2

32. Which statement is true regarding gases when compared to liquids?

A. Gases have lower compressibility and higher density

B. Gases have lower compressibility and lower density

C. Gases have higher compressibility and higher density

D. Gases have higher compressibility and lower density

E. None of the above

33. When nonvolatile solute molecules are added to a solution, the vapor pressure of the solution:

A. stays the same

B. increases

C. decreases

D. is directly proportional to the second power of the amount added

E. is directly proportional to the square root of the amount added

34. Which of the following laws states that the pressure exerted by a mixture of gases equals the sum of the individual gas pressures?

A. Gay-Lussac's law

B. Dalton's law

C. Charles's law

D. Boyle's law

E. Avogadro's law

35. Which characteristics best describe a solid?

   A. Definite volume; the shape of the container; no intermolecular attractions

   B. Volume and shape of the container; no intermolecular attractions

   C. Definite shape and volume; strong intermolecular attractions

   D. Definite volume; the shape of the container; moderate intermolecular attractions

   E. Volume and shape of the container; strong intermolecular attractions

36. Which of the following statements is true if three 2.0 L flasks are filled with H_2, O_2 and He, respectively, at STP?

   A. There are twice as many He atoms as H_2 or O_2 molecules

   B. There are four times as many H_2 or O_2 molecules as He atoms

   C. Each flask contains the same number of atoms

   D. There are twice as many H_2 or O_2 molecules as He atoms

   E. The number of H_2 or O_2 molecules is the same as the number of He atoms

37. Which of the following atoms could interact through a hydrogen bond?

   A. The hydrogen of an amine and the oxygen of an alcohol

   B. The hydrogen on an aromatic ring and the oxygen of carbon dioxide

   C. The oxygen of a ketone and the hydrogen of an aldehyde

   D. The oxygen of methanol and hydrogen on the methyl carbon of methanol

   E. None of the above

38. Which of the following laws states that the pressure and volume are inversely proportional for gas at a constant temperature?

   A. Dalton's law   **C.** Boyle's law

   B. Gay-Lussac's law   **D.** Charles's law

                                   E. Avogadro's law

39. As an automobile travels the highway, the temperature of the air inside the tires [   ] and the pressure [   ]?

   A. increases … decreases   **C.** decreases … decreases

   B. increases … increases   **D.** decreases … increases

                                    E. none of the above

40. Using the following unbalanced chemical reaction, what volume of H_2 gas at 780 mmHg and 23 °C is required to produce 12.5 L of NH_3 gas at the same temperature and pressure?

$$N_2\ (g) + H_2\ (g) \rightarrow NH_3\ (g)$$

A. 21.4 L

B. 15.0 L

C. 18.8 L

D. 13.0 L

E. 12.5 L

Notes for active learning

Practice Set 3: Questions 41–60

41. A mixture of gases containing 16 g of O_2, 14 g of N_2 and 88 g of CO_2 is collected above water at a temperature of 23 °C. The pressure is 1 atm, and the vapor pressure of water is 38 torr. What is the partial pressure exerted by CO_2?

A. 283 torr

B. 367 torr

C. 481 torr

D. 549 torr

E. 583 torr

42. Which of the following demonstrate colligative properties?

    I. Freezing point

    II. Boiling point

    III. Vapor pressure

A. I only

B. II only

C. III only

D. I and II only

E. I, II and III

43. What is the proportionality relationship between the pressure of a gas and its volume?

A. directly

B. inversely

C. pressure is raised to the 2nd power

D. pressure raised to the $\sqrt{2}$ power

E. none of the above

44. Which of the following statements about gases is correct?

A. Formation of homogeneous mixtures, regardless of the non-reacting gas components

B. Relatively long distances between molecules

C. High compressibility

D. No attractive forces between gas molecules

E. All of the above

45. Which of the following describes a substance in the solid physical state?

> I. It compresses negligibly
> II. It has a fixed volume
> III. It has a fixed shape

A. I only

B. II only

C. I and III only

D. II and III only

E. I, II and III

46. Which of the following is NOT a unit used in measuring pressure?

A. kilometers Hg

B. millimeters Hg

C. atmosphere

D. Pascal

E. torr

47. Which of the following compounds has the highest boiling point?

A. CH_4

B. $CHCl_3$

C. CH_3COOH

D. NH_3

E. CH_2Cl_2

48. Identify the decreasing ordering of attractions among particles in the three states of matter.

A. gas > liquid > solid

B. gas > solid > liquid

C. solid > liquid > gas

D. liquid > solid > gas

E. solid > gas > liquid

49. A sample of N_2 gas occupies a volume of 190 mL at STP. What volume will it occupy at 660 mmHg and 295 K?

A. 1.15 L

B. 0.760 L

C. 0.214 L

D. 1.84 L

E. 0.197 L

50. A nonvolatile liquid would have:

   A. a highly explosive propensity

   B. strong attractive forces between molecules

   C. weak attractive forces between molecules

   D. a high vapor pressure at room temperature

   E. weak attractive forces within molecules

51. A closed-end manometer was constructed from a U-shaped glass tube. It was loaded with mercury so that the closed side was filled to the top, which was 820 mm above the neck, while the open end was 160 mm above the neck. The manometer was taken into a chamber used for training astronauts. What is the highest pressure that can be read with assurance on this manometer?

   A. 66.0 torr

   B. 660 torr

   C. 220 torr

   D. 760 torr

   E. 5.13 torr

52. Vessels X and Y each contain 1.00 L of a gas at STP, but vessel X contains oxygen, while vessel Y contains nitrogen. Assuming the gases behave as ideal, they have the same:

   I. number of molecules   II. density   III. kinetic energy

   A. II only

   B. III only

   C. I and III only

   D. I, II and III

   E. I only

53. What condition must be satisfied for the noble gas Xe to exist in the liquid phase at 180 K, which is a temperature significantly greater than its normal boiling point?

   A. external pressure > vapor pressure of xenon

   B. external pressure < vapor pressure of xenon

   C. external pressure = partial pressure of water

   D. temperature is increased quickly

   E. temperature is increased slowly

54. Matter is nearly incompressible in which of these states?

      I. solid   II. Liquid   III. gas

A. I only

B. II only

C. III only

D. I and II only

E. I, II and III

55. Which of the following laws states that volume and temperature are directly proportional for a gas at constant pressure?

A. Gay-Lussac's law

B. Dalton's law

C. Charles' law

D. Boyle's law

E. Avogadro's law

56. Which transformation describes sublimation?

A. solid $\rightarrow$ liquid

B. solid $\rightarrow$ gas

C. liquid $\rightarrow$ solid

D. liquid $\rightarrow$ gas

E. gas $\rightarrow$ liquid

57. What is the ratio of the diffusion rate of O_2 molecules to the diffusion rate of H_2 molecules if six moles of O_2 gas and six moles of H_2 gas are placed in a large vessel, and the gases and vessels are at the same temperature?

A. 4:1

B. 1:4

C. 12:1

D. 1:1

E. 2:1

58. How many molecules of neon gas are present in 6 liters at 10 °C and 320 mmHg? (Use the ideal gas constant R = 0.0821 L·atm K^{-1} mol^{-1})

A. (320 mmHg / 760 atm)·(0.821)·(6 L) / (6 × 10^{23})·(283 K)

B. (320 mmHg / 760 atm)·(6 L)·(6 × 10^{23}) / (0.0821)·(283 K)

C. (320 mmHg)·(6 L)·(283 K)·(6 × 10^{23})

D. (320 mmHg / 760 atm)·(6 L)·(283 K)·(6 × 10^{23}) / (0.821)

E. (320 mmHg / 283 K)·(6 L)·(6 × 10^{23}) / (0.0821)·(760 atm)

59. Which gas has the greatest density at STP:

    A. CO_2

    B. O_2

    C. N_2

    D. NO

    E. more than one

60. A hydrogen bond is a special type of:

    A. dipole–dipole attraction involving hydrogen bonded to another hydrogen atom

    B. attraction involving any molecules that contain hydrogens

    C. dipole-dipole attraction involving hydrogen bonded to a highly electronegative atom

    D. dipole–dipole attraction involving hydrogen bonded to any other atom

    E. London dispersion force involving hydrogen bonded to an electropositive atom

Notes for active learning

Notes for active learning

Practice Set 4: Questions 61–80

61. What is the mole fraction of H_2 in a gaseous mixture that consists of 9.50 g of H_2 and 14.0 g of Ne in a 4.50-liter container maintained at 37.5 °C?

   A. 0.13   **C.** 0.67
   B. 0.43   **D.** 0.87
                               E. 1.2

62. Which of the following is true about Liquid A if the vapor pressure of Liquid A is greater than Liquid B?

   A. Liquid A has a higher heat of fusion
   B. Liquid A has a higher heat of vaporization
   C. Liquid A forms stronger bonds
   D. Liquid A boils at a higher temperature
   E. Liquid A boils at a lower temperature

63. When 25 g of non-ionizable compound X is dissolved in 1 kg of camphor, the freezing point of the camphor falls 2.0 K. What is the approximate molecular weight of compound X? (Use $K_{camphor} = 40$)

   A. 50 g/mol   **C.** 5,000 g/mol
   B. 500 g/mol   **D.** 5,500 g/mol
                                 E. 5,750 g/mol

64. The boiling point of a liquid is the temperature:

   A. where sublimation occurs
   B. where the vapor pressure of the liquid is less than the atmospheric pressure over the liquid
   C. equal to or greater than 100 °C
   D. where the rate of sublimation equals evaporation
   E. where the vapor pressure of the liquid equals the atmospheric pressure over the liquid

65. The van der Waals equation $[(P + n^2a / v^2) \cdot (V - nb) = nRT]$ is used to describe nonideal gases. The terms n^2a / v^2 and nb stand for, respectively:

    A. volume of gas molecules and intermolecular forces

    B. nonrandom movement and intermolecular forces between gas molecules

    C. nonelastic collisions and volume of gas molecules

    D. intermolecular forces and volume of gas molecules

    E. nonrandom movement and volume of gas molecules

66. What are the units of the gas constant R?

    A. atm·K/L·mol

    B. atm·K/mol

    C. mol·L/atm·K

    D. mol·K/L·atm

    E. L·atm/mol·K

67. A sample of a gas occupying a volume of 120.0 mL at STP was placed in a different vessel with a volume of 155.0 mL, in which the pressure was measured at 0.80 atm. What was its temperature?

    A. 9.1 °C

    B. 43.1 °C

    C. 4.1 °C

    D. 93.6 °C

    E. 108.3 °C

68. Why does a beaker of water begin to boil at 22 °C when placed in a closed chamber, and a vacuum pump is used to evacuate the air from the chamber?

    A. The vapor pressure decreases

    B. Air is released from the water

    C. The atmospheric pressure decreases

    D. The vapor pressure increases

    E. The atmospheric pressure increases

69. According to the kinetic theory, what happens to the kinetic energy of gaseous molecules when the temperature of a gas decreases?

    A. Increase as does velocity

    B. Remains constant, as does velocity

    C. Increases and velocity decreases

    D. Decreases and velocity increases

    E. Decreases as does velocity

70. A sample of SO_3 gas is decomposed into SO_2 and O_2.

$$2\ SO_3\ (g) \rightarrow 2\ SO_2\ (g) + O_2\ (g)$$

If the pressure of SO_2 and O_2 is 1,250 torrs, what is the partial pressure of O_2 in torr?

A. 417 torr

B. 1,040 torr

C. 1,250 torr

D. 884 torr

E. 12.50 torr

71. For the balanced reaction $2\ Na + Cl_2 \rightarrow 2\ NaCl$, which of the following is solid?

    I. Na   II. Cl   III. NaCl

A. I only

B. II only

C. III only

D. I and III only

E. I, II and III

72. According to Charles's law, what happens to gas as temperature increases?

A. volume decreases

B. volume increases

C. pressure decreases

D. pressure increases

E. mole fraction increases

73. How does the pressure of a sample of gas change if the moles of gas remain constant while the volume is halved and the temperature is quadrupled?

A. decrease by a factor of 8

B. quadruple

C. decrease by a factor of 4

D. decrease by a factor of 2

E. increase by a factor of 8

74. For a fixed quantity of gas, gas laws describe the relationships between pressure and which two variables?

A. chemical identity; mass

B. volume; chemical identity

C. temperature; volume

D. temperature; size

E. volume; size

75. What is the term for a direct change of state from a solid to a gas?

A. sublimation

B. vaporization

C. condensation

D. deposition

E. melting

76. The average speed at which a methane molecule effuses at 28.5 °C is 631 m/s. The average speed at which a krypton molecule effuses at the same temperature is:

A. 123 m/s

B. 276 m/s

C. 312 m/s

D. 421 m/s

E. 633 m/s

77. A chemical reaction A $(s) \rightarrow$ B (s) + C (g) occurs when substance A is vigorously heated. The molecular mass of the gaseous product was determined from the following experimental data:

Mass of A before reaction: 5.2 g

Mass of A after reaction: 0 g

Mass of residue B after cooling and weighing when no more gas evolved: 3.8 g

When all of the gas C evolved, it was collected and stored in a 668.5 mL glass vessel at 32.0 °C, and the gas exerted a pressure of 745.5 torrs.

Use the ideal gas constant R equals 0.0821 L·atm K^{-1} mol^{-1}

From this data, determine the apparent molecular mass of *C*, assuming it behaves as an ideal gas:

A. 6.46 g/mol

B. 46.3 g/mol

C. 53.9 g/mol

D. 72.2 g/mol

E. 142.7 g/mol

78. Which of the following describes a substance in the liquid physical state?

    I. It has a variable shape

    II. It compresses negligibly

    III. It has a fixed volume

A. I only

B. II only

C. I and III only

D. II and III only

E. I, II and III

79. Which is the strongest form of intermolecular attraction between water molecules?

A. ion-dipole

B. covalent bonding

C. induced dipole-induced dipole

D. hydrogen bonding

E. polar-induced dipolar

80. When a helium balloon is placed in a freezer, the temperature in the balloon [   ] and the volume [   ]?

A. increases … increases

B. increases … decreases

C. decreases … increases

D. decreases … decreases

E. none of the above

Notes for active learning

Practice Set 5: Questions 81–100

81. How does the volume of a fixed sample of gas change if the pressure is doubled?

A. Decreases by a factor of 2

B. Increases by a factor of 4

C. Doubles

D. Remains the same

E. Requires more information

82. What is the term that refers to the frequency and energy of gas molecules colliding with the walls of the container?

I. partial pressure II. vapor pressure III. gas pressure

A. I only

B. II only

C. III only

D. I and II only

E. I, II and III

83. The freezing point changes by 10 K when an unknown amount of toluene is added to 100 g of benzene. Find the number of moles of toluene added from the given data. (Use the $K_{benzene} = 5.0$ and $K_{toluene} = 8.4$)

A. 0.14 B. 0.20 C. 0.23 D. 0.27 E. 0.31

84. A container is labeled "Ne, 5.0 moles" but has no pressure gauge. By measuring the temperature and determining the volume of the container, a chemist uses the ideal gas law to estimate the pressure inside the container. If the container was mislabeled and contained 5.0 moles of He, not Ne, how would this affect the scientist's estimate?

A. The estimate is correct, but only because both gases are monatomic

B. The estimate is too high

C. The estimate is slightly too low

D. The estimate is significantly lower because of the large difference in molecular mass

E. The estimate is correct because the identity of the gas is irrelevant

85. According to the kinetic theory, which is NOT true of ideal gases?

  A. For a sample of gas molecules, average kinetic energy is directly proportional to temperature

  B. There are no attractive or repulsive forces between gas molecules

  C. Collisions among gas molecules are perfectly elastic

  D. There is no transfer of kinetic energy during collisions between gas molecules

  E. None of the above

86. At what temperature is degrees Celsius equivalent to degrees Fahrenheit?

  A. 0   **B.** −10   **C.** −25   **D.** −40   **E.** −50

87. Which statement about the boiling point of water is NOT correct?

  A. At sea level and a pressure of 760 mmHg, the boiling point is 100 °C

  B. In a pressure cooker, shorter cooking times are achieved due to the change in boiling point

  C. The boiling point is greater than 100 °C in a pressure cooker

  D. The boiling point is less than 100 °C for locations at low elevations

  E. The boiling point is greater than 100 °C for locations at low elevations

88. Which of the following compounds has the lowest boiling point?

  A. CH_4 **C.** CH_3CH_2OH

  B. $CHCl_3$ **D.** NH_3

                                  E. CH_2Cl_2

89. As the pressure is increased on solid CO_2, the melting point is:

  A. decreased   **C.** increased

  B. unchanged   **D.** inversely proportional to the square

  root of the change   **E.** cannot be determined without

  further information

90. A vessel contains 32.0 g of CH_4 gas and 12.75 g of NH_3 gas at a combined pressure of 2.4 atm. What is the partial pressure of the NH_3 gas?

  A. 0.30 atm   **C.** 0.44 atm

  B. 1.22 atm   **D.** 1.88 atm

                                  E. 0.66 atm

91. Which of the following statements bests describes a liquid?

   A. Definite shape, but an indefinite volume

   B. Indefinite shape, but a definite volume

   C. Indefinite shape and volume

   D. Definite shape and volume

   E. Definite shape, but indefinite mass

92. Assuming constant pressure, if a volume of nitrogen gas at 420 K decreases from 100 mL to 50 mL, what is the final temperature in Kelvin?

   A. 630 K

   B. 420 K

   C. 910 K

   D. 150 K

   E. 210 K

93. Which of the following laws states that pressure and Kelvin temperature are directly proportional for a gas at constant volume?

   A. Gay-Lussac's law

   B. Dalton's law

   C. Charles' law

   D. Boyle's law

   E. Avogadro's law

94. The Gay–Lussac's law of increased pressure due to increased temperature is explained using Kinetic Molecular Theory stating that the pressure must increase because the molecules:

   A. increase in size

   B. move slower

   C. decrease in size

   D. strike the container walls less often

   E. strike the container walls more often

95. Which of the following terms does NOT involve the solid state?

   A. solidification

   B. sublimation

   C. evaporation

   D. melting

   E. deposition

96. Which of the following compounds exhibits primarily dipole-dipole intermolecular forces?

   A. CO_2 **B.** F_2 **C.** CH_3-O-CH_3 **D.** CH_3CH_3 **E.** CCl_4

97. Which of the following increases the pressure of a gas?

A. Decreasing the volume

B. Increasing the number of molecules

C. Increasing temperature

D. None of the above

E. All of the above

98. According to Avogadro's law, the volume of a gas [   ] as the [   ] increases while [   ] is held constant.

A. increases… temperature… pressure and number of moles

B. decreases… pressure… temperature and number of moles

C. increases… pressure… temperature and number of moles

D. decreases… number of moles… pressure and temperature

E. increases… number of moles… pressure and temperature

99. A gas initially filled a 3.0 L container. Heat was then added to the gas, which raised its temperature from 100 K to 150 K while increasing its pressure from 3.0 atm to 4.5 atm. What is the new volume of the gas?

A. 1.4 L   **B.** 3.0 L   **C.** 2.0 L   **D.** 4.5 L   **E.** 6.0 L

100. Consider a 10.0 liters sample of helium and a 10.0 liters sample of neon, both at 23 °C, 2.0 atm. Which statement regarding these samples is NOT true?

A. The density of the neon sample is greater than the density of the helium sample

B. Each sample contains the same number of moles of gas

C. Each sample weighs the same amount

D. Each sample contains the same number of atoms of gas

E. All statements are true

Answer Key & Detailed Explanations

Answer Key

1: A	11: B	21: E	31: B	41: C	51: B	61: D	71: D	81: A	91: B
2: C	12: C	22: A	32: D	42: E	52: C	62: E	72: B	82: C	92: E
3: C	13: A	23: E	33: C	43: B	53: A	63: B	73: E	83: B	93: A
4: A	14: D	24: D	34: B	44: E	54: D	64: E	74: C	84: E	94: E
5: A	15: C	25: D	35: C	45: E	55: C	65: D	75: A	85: D	95: C
6: B	16: A	26: A	36: E	46: A	56: B	66: E	76: B	86: D	96: C
7: B	17: C	27: D	37: A	47: C	57: B	67: A	77: C	87: D	97: E
8: B	18: C	28: B	38: C	48: C	58: B	68: C	78: E	88: A	98: E
9: E	19: C	29: E	39: B	49: C	59: A	69: E	79: D	89: C	99: B
10: D	20: E	30: A	40: C	50: B	60: C	70: A	80: D	90: E	100: C

Practice Set 1: Questions 1–20

1. A is correct.

Solids, liquids, and gases have a vapor pressure that increases from solid to gas.

Vapor pressure is the pressure exerted by a vapor in equilibrium with its condensed phases (i.e., solid or liquid) in a closed system at a given temperature.

Vapor pressure is a colligative property of a substance and depends only on the number of solutes present, not on their identity.

2. C is correct.

Charles' law (i.e., the law of volumes) explains how, at constant pressure, gases behave when the temperature changes:

$$V \propto T$$

or

$$V / T = \text{constant}$$

or

$$(V_1 / T_1) = (V_2 / T_2)$$

Volume and temperature are proportional. Doubling the temperature at constant pressure doubles the volume.

3. C is correct.

Colligative properties include lowering of vapor pressure, the elevation of boiling point, depression of freezing point, and increased osmotic pressure.

The addition of solute to a pure solvent lowers the vapor pressure of the solvent; therefore, a higher temperature is required to bring the vapor pressure of the solution in an open container up to the atmospheric pressure. This increases the boiling point.

Because adding solute lowers the vapor pressure, the solution's freezing point decreases (e.g., automobile antifreeze).

4. A is correct.

At a pressure and temperature corresponding to the triple point (point D on the graph) of a substance, all three states (gas, liquid and solid) exist in equilibrium.

The critical point (point E on the graph) is the endpoint of the phase equilibrium curve where the liquid and its vapor become indistinguishable.

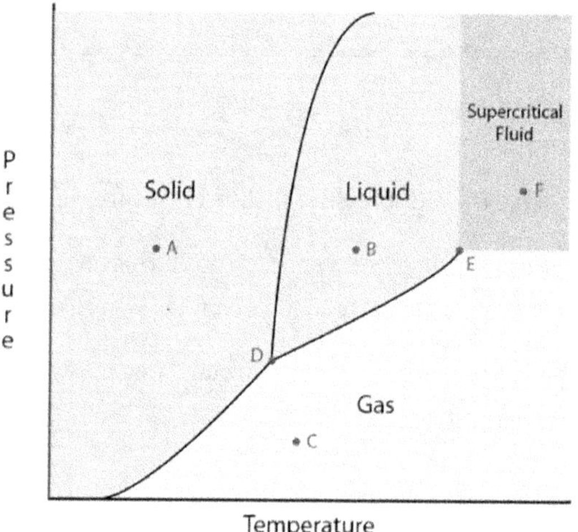

Phase diagram of pressure vs. temperature

5. A is correct.

In the van der Waals equation, a is the negative deviation due to attractive forces, and b is the positive deviation due to molecular volume.

6. B is correct.

R is the symbol for the ideal gas constant.

R is expressed in (L × atm) / mole × K and the value = 0.0821.

Convert to different units (torr and mL)

$$0.0821[(L \times atm) / (mole \times K)] \times (760 \text{ torr/atm}) \times (1{,}000 \text{ mL/L})$$

$$R = 62{,}396 \text{ (torr} \times \text{mL) / mole} \times K$$

7. B is correct.

Ideal gas law:

$$PV = nRT$$

where P is pressure, V is volume, n is the number of molecules, R is the ideal gas constant, and T is the temperature of the gas

R and n are constant.

From this equation, if both pressure and temperature are halved, there would be no effect on the volume.

8. B is correct.

Kinetic Molecular Theory of gas molecules states that the average kinetic energy per molecule in a system is proportional to the temperature of the gas.

Since it is given that containers X and Y are at the same temperature and pressure, then molecules of both gases must possess the same amount of average kinetic energy.

9. E is correct.

Barometers and manometers are used to measure pressure.

Barometers are designed to measure atmospheric pressure, while a manometer can measure the pressure that is lower than atmospheric pressure.

A manometer has both ends of the tube open to the outside (while some may have one end closed), whereas a barometer is a type of closed-end manometer with one end of the glass tube closed and sealed with a vacuum.

The atmospheric pressure is 760 mmHg, so the barometer should accommodate that.

10. D is correct.

Vapor pressure is the pressure exerted by a vapor in equilibrium with its condensed phases (i.e., solid or liquid) in a closed system at a given temperature.

Raoult's law states that the partial vapor pressure of each component of an ideal mixture of liquids equals the vapor pressure of the pure component multiplied by its mole fraction in the mixture.

In exothermic reactions, the vapor pressure deviates negatively from Raoult's law.

Depending on the ratios of the liquids in a solution, the vapor pressure could be lower than either or just lower than X because X is a higher boiling point, thus a lower vapor pressure.

The boiling point increases from adding Y to the mixture because the vapor pressure decreases.

11. B is correct.

$$2\,Na\,(s) + Cl_2\,(g) \rightarrow 2\,NaCl\,(s)$$

In its elemental form, chlorine exists as a gas.
In its elemental form, Na exists as a solid.

12. C is correct.

The molecules of an ideal gas do not occupy a significant amount of space and exert no intermolecular forces. The molecules of a real gas do occupy space and do exert (weak attractive) intermolecular forces.

However, both an ideal gas and a real gas have pressure, which is created from molecular collisions with the walls of the container.

13. A is correct.

Boyle's law (i.e., pressure-volume law) states that pressure and volume are inversely proportional:

$$(P_1 V_1) = (P_2 V_2)$$

Solve for the final pressure:

$$P_2 = (P_1 V_1) / V_2$$

$$P_2 = [(0.950\ atm) \times (2.75\ L)] / (0.450\ L)$$

$$P_2 = 5.80\ atm$$

14. D is correct.

Avogadro's law is an experimental gas law relating the volume of gas to the amount of substance of gas present.

Avogadro's law states that equal volumes of gases at the same temperature and pressure have the same number of molecules.

For a given mass of an ideal gas, the volume and amount (i.e., moles) of the gas are directly proportional if the temperature and pressure are constant:

$$V \alpha n$$

where V = volume and n = number of moles of the gas.

Charles' law (i.e., the law of volumes) explains how, at constant pressure, gases behave when the temperature changes:

$$V \alpha T$$

or

$$V / T = \text{constant}$$

or

$$(V_1 / T_1) = (V_2 / T_2)$$

Volume and temperature are proportional. Therefore, an increase in one term increases the other.

Gay-Lussac's law (i.e., pressure-temperature law) states that pressure is proportional to temperature:

$$P \alpha T$$

or $\quad (P_1 / T_1) = (P_2 / T_2)$

or $\quad (P_1 T_2) = (P_2 T_1)$

or $\quad P / T = \text{constant}$

If the pressure of a gas increases, the temperature increases.

Boyle's law (i.e., pressure-volume law) states that pressure and volume are inversely proportional:

15. C is correct.

Ideal gas law: $PV = nRT$

where P is pressure, V is volume, n is the number of molecules, R is the ideal gas constant, and T is the temperature of the gas.

R and n are constant.

If T is constant, the equation becomes $PV = $ constant.

They are inversely proportional: if one of the values is reduced, the other increases.

16. A is correct.

Vaporization refers to the change of state from a liquid to a gas. There are two types of vaporization: boiling and evaporation differentiated based on the temperature they occur.

Evaporation occurs at a temperature below the boiling point, while boiling occurs at or above the boiling point.

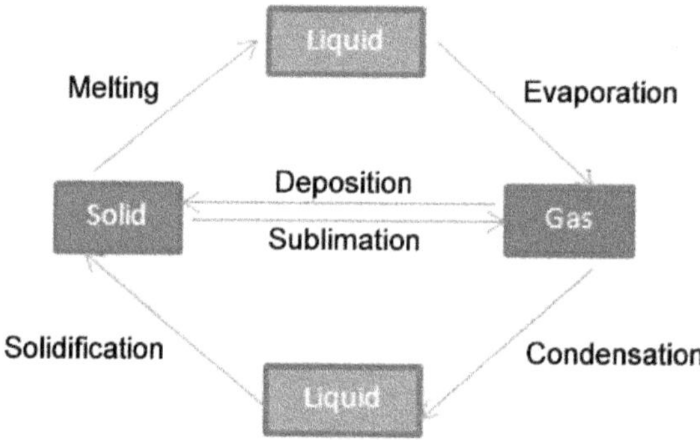

Interconversion of states of matter

17. C is correct.

Ideal gas law:

$$PV = nRT$$

from which simpler gas laws such as Boyle's, Charles', and Avogadro's laws are derived.

The value of n and R are constant.

The common format of the combined gas law:

$$(P_1V_1) / T_1 = (P_2V_2) / T_2$$

Try modifying the equations to recreate the formats of the equation provided by the problem.

$T_2 = T_1 \times P_1 / P_2 \times V_2 / V_1$ should be written as $T_2 = T_1 \times V_1/V_2 \times P_1/P_2$

18. C is correct.

Intermolecular forces act between neighboring molecules. Examples include hydrogen bonding, dipole-dipole, dipole-induced dipole, and van der Waals (i.e., London dispersion) forces.

A dipole-dipole attraction involves asymmetric, polar molecules (based on differences in electronegativity between atoms) that create a dipole moment (i.e., a net vector indicating the force).

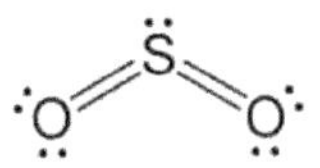

Sulfur dioxide with indicated bond angle

19. C is correct.

Increasing the pressure of the gas above the liquid puts stress on the equilibrium of the system. Gas molecules start to collide with the liquid surface more often, which increases the rate of gas molecules entering the solution, thus increasing the solubility.

20. E is correct.

Avogadro's law states the correlation between volume and moles (n).

Avogadro's law is an experimental gas law relating the volume of a gas to the amount of gas present.

At the same pressure and temperature, equal volumes of gases have the same number of molecules.

$$V \propto n$$

or

$$V / n = k$$

where V is the volume of the gas, n is the number of moles of the gas, and k is a constant equal to RT/P (where R is the universal gas constant, T is the temperature in Kelvin and P is the pressure).

For comparing the same substance under two sets of conditions, the law is expressed as:

$$V_1 / n_1 = V_2 / n_2$$

Notes for active learning

Practice Set 2: Questions 21–40

21. E is correct.

Colligative properties of solutions depend on the ratio of the number of solute particles to the number of solvent molecules in a solution and not on the type of chemical species present.

Colligative properties include: lowering of vapor pressure, the elevation of boiling point, depression of freezing point, and increased osmotic pressure.

Boiling point (BP) elevation: $\Delta BP = iKm$

where i = the number of particles produced when the solute dissociates, K = boiling elevation constant and m = molality (moles/kg solvent).

In this problem, acid molality is not known.

22. A is correct.

An ideal gas has no intermolecular forces, indicating that its molecules have no attraction to each other. The molecules of a real gas, however, do have intermolecular forces, although these forces are extremely weak.

Therefore, the molecules of a real gas are slightly attracted to one another, although the attraction is nowhere near as strong as the attraction in liquids and solids.

23. E is correct.

Standard temperature and pressure (STP) has a temperature of 273.15 K (0 °C, 32 °F) and a pressure of 10^5 Pa (100 kPa, 750.06 mmHg, 1 bar, 14.504 psi, 0.98692 atm).

The mm of Hg was defined as the pressure generated by a column of mercury one millimeter high. The pressure of mercury depends on temperature and gravity.

This variation in mmHg and torr is a difference in units of about 0.000015%.

In general,

      1 torr = 1 mm of Hg = 0.0013158 atm.

      750.06 mmHg = 0.98692 atm.

24. D is correct.

At lower temperatures, the potential energy due to the intermolecular forces are more significant compared to the kinetic energy; this causes the pressure to be reduced because the gas molecules are attracted to each other.

25. D is correct.

Dalton's law (i.e., the law of partial pressures) states that *the pressure exerted by a mixture of gases is equal to the sum of the individual gas pressures.*

The pressure due to N_2 and CO_2:

$$(320 \text{ torr} + 240 \text{ torr}) = 560 \text{ torr}$$

The partial pressure of O_2 is:

$$740 \text{ torr} - 560 \text{ torr} = 180 \text{ torr}$$

$$180 \text{ torr} / 740 \text{ torr} = 24\%$$

26. A is correct.

Ideal gas law:

$$PV = nRT$$

where P is pressure, V is volume, n is the number of molecules, R is the ideal gas constant, and T is the temperature of the gas.

If the volume is reduced by ½, the number of moles is reduced by ½.

Pressure is reduced to 90%, so the number of moles is reduced by 90%.

Therefore, the total reduction in moles is ($½ \times 90\%$) = 45%.

Mass is proportional to the number of moles for a given gas so that the mass reduction can be calculated directly:

$$\text{New mass is } 45\% \text{ of } 40 \text{ grams} = (0.45 \times 40 \text{ g}) = 18 \text{ grams}$$

27. D is correct.

Evaporation describes the phase change from liquid to gas.

The mass of the molecules and the attraction of the molecules with their neighbors (to form intermolecular attractions) determine their kinetic energy.

The increase in kinetic energy is required for individual molecules to move from the liquid to the gaseous phase.

28. B is correct.

The molecules of an ideal gas exert no attractive forces.

Therefore, a real gas behaves most nearly like an ideal gas when it is at high temperature and low pressure because under these conditions, the molecules are far apart from each other and exert little or no attractive forces on each other.

29. E is correct.

Hydroxyl (~OH) groups greatly increase the boiling point because they form hydrogen bonds with ~OH groups of neighboring molecules.

Hydrocarbons are nonpolar molecules, which means that the dominant intermolecular force is London dispersion. This force gets stronger as the number of atoms in each molecule increases. The stronger force increases the boiling point.

The branching of the hydrocarbon affects the boiling point. Straight molecules have slightly higher boiling points than branched molecules with the same number of atoms. The reason is that straight molecules can align parallel against each other, and atoms in the molecules are involved in the London dispersion forces.

Another factor is the presence of other heteroatoms (i.e., atoms other than carbon and hydrogen). For example, the electronegative oxygen atom between carbon groups or in an ether (C–O–C) slightly increases the boiling point.

30. A is correct.

Kinetic theory explains the macroscopic properties of gases (e.g., temperature, volume, and pressure) by their molecular composition and motion.

The gas pressure is due to the collisions on the walls of a container from molecules moving at different velocities.

Temperature $= \frac{1}{2}mv^2$

31. B is correct.

Methanol (CH_3OH) is alcohol that participates in hydrogen bonding.

Therefore, this gas experiences the strongest intermolecular forces.

32. D is correct.

Density = mass / volume

Gas molecules have a large amount of space between them; therefore, they can be pushed together, and thus gases are very compressible. Because there is such a large amount of space between each molecule in a gas, the extent to which the gas molecules can be pushed together is much greater than the extent to which liquid molecules can be pushed together. Therefore, gases have greater compressibility than liquids.

Gas molecules are further apart than liquid molecules, which is why gases have a smaller density.

33. C is correct.

Vapor pressure is the pressure exerted by a vapor in equilibrium with its condensed phases (i.e., solid or liquid) in a closed system at a given temperature.

Vapor pressure is inversely correlated with the strength of the intermolecular force.

The molecules are more likely to stick together in the liquid form with stronger intermolecular forces, and fewer participate in the liquid-vapor equilibrium.

The vapor pressure of a liquid decreases when a nonvolatile substance is dissolved into a liquid.

The decrease in the vapor pressure of a substance is proportional to the number of moles of the solute dissolved in a definite weight of the solvent. This is Raoult's law.

34. B is correct.

Dalton's law (i.e., the law of partial pressures) states that *the pressure exerted by a mixture of gases is equal to the sum of the individual gas pressures*. It is an empirical law that was observed by English chemist John Dalton and is related to the ideal gas laws.

35. C is correct.

Solids have a definite shape and volume. For example, a granite block does not change its shape or its volume regardless of the container in which it is placed.

Molecules in a solid are very tightly packed due to the strong intermolecular attractions, which prevent the molecules from moving around.

36. E is correct.

Ideal gas law: $PV = nRT$

where P is pressure, V is volume, n is the number of molecules, R is the ideal gas constant, and T is the temperature of the gas.

At STP (standard conditions for temperature and pressure), the pressure and temperature for the three flasks are the same. It is known that the volume is the same in each case – 2.0 L.

Therefore, since R is a constant, the number of molecules "n" must be the same for the ideal gas law to hold.

37. A is correct.

Hydrogens, bonded directly to F, O or N, participate in hydrogen bonds. The hydrogen is partially positive (i.e., delta plus: $\partial+$) due to the bond to these electronegative atoms. The lone pair of electrons on the F, O or N interacts with the partial positive ($\partial+$) hydrogen to form a hydrogen bond.

D: the hydrogen on the methyl carbon and that carbon is not attached to N, O, or F. Therefore, that hydrogen cannot form a hydrogen bond, even though there is available oxygen on the other methanol to form a hydrogen bond.

38. C is correct.

Boyle's law (Mariotte's law or the Boyle-Mariotte law) is an experimental gas law that describes how the volume of a gas increases as the pressure decreases (i.e., they are inversely proportional) if the temperature is constant.

Boyle's law (i.e., pressure-volume law) states that pressure and volume are inversely proportional:

$$P_1V_1 = P_2V_2$$

or　　　$P \times V = \text{constant}$

If the volume of a gas increases, its pressure decreases proportionally.

Dalton's law (i.e., the law of partial pressures) states that *the pressure exerted by a mixture of gases is equal to the sum of the individual gas pressures.*

Charles' law (i.e., the law of volumes) explains how, at constant pressure, gases behave when the temperature changes:

$$V \, \alpha \, T$$

or　　　$V / T = \text{constant}$

or　　　$(V_1 / T_1) = (V_2 / T_2)$

Volume and temperature are proportional: an increase in one results in an increase in the other.

Gay-Lussac's law (i.e., pressure-temperature law) states that pressure is proportional to temperature:

$$P \propto T$$

or $\quad (P_1 / T_1) = (P_2 / T_2)$

or $\quad (P_1 T_2) = (P_2 T_1)$

or $\quad P / T = \text{constant}$

If the pressure of a gas increases, the temperature increases.

Avogadro's law is an experimental gas law relating the volume of a gas to the amount of substance of gas present. It states that *equal volumes of all gases, at the same temperature and pressure, have the same number of molecules.*

39. B is correct.

As the automobile travels the highway, friction is generated between the road and its tires. The heat energy increases the temperature of the air inside the tires, causing the molecules to have more velocity. These fast-moving molecules collide with the walls of the tire at a higher rate, and thus the pressure is increased.

Gay-Lussac's law (i.e., pressure-temperature law) states that pressure is proportional to temperature:

$$P \propto T$$

or $\quad (P_1 / T_1) = (P_2 / T_2)$

or $\quad (P_1 T_2) = (P_2 T_1)$

or $\quad P / T = \text{constant}$

If the temperature of a gas is increased, the pressure increases.

40. C is correct.

The balanced chemical equation:

$$N_2 + 3\ H_2 \rightarrow 2\ NH_3$$

Use the balanced coefficients from the written equation, apply dimensional analysis to solve for the volume of H_2 needed to produce 12.5 L NH_3:

$$V_{H2} = V_{NH3} \times (mol\ H_2\ /\ mol\ NH_3)$$

$$V_{H2} = (12.5\ L) \times (3\ mol\ /\ 2\ mol)$$

$$V_{H2} = 18.8\ L$$

Notes for active learning

Practice Set 3: Questions 41–60

41. C is correct.

Dalton's law (i.e., the law of partial pressures) states that *the pressure exerted by a mixture of gases is equal to the sum of the individual gas pressures*.

Convert the masses of the gases into moles:

Moles of O_2:   16 g of O_2 ÷ 32 g/mole = 0.5 mole

Moles of N_2:   14 g of N_2 ÷ 28 g/mole = 0.5 mole

Mole of CO_2:   88 g of CO_2 ÷ 44 g/mole = 2 moles

Total moles: (0.5 mol + 0.5 mole + 2 mol) = 3 moles

The pressure of 1 atm (or 760 mmHg) has 38 mmHg (1 torr = 1 mmHg) contributed as H_2O vapor.

The partial pressure of CO_2:

mole fraction × (total pressure of the gas mixture – H_2O vapor)

(2 moles CO_2 / 3 moles total gas)] × (760 mmHg – 38 mmHg)

partial pressure of CO_2 = 481 mmHg

42. E is correct.

Colligative properties are properties of solutions that depend on the ratio of the number of solute particles to the number of solvent molecules in a solution, not on the type of chemical species present.

Colligative properties include: lowering of vapor pressure, the elevation of boiling point, depression of freezing point, and increased osmotic pressure

Dissolving a solute into a solvent alters the solvent's freezing point, melting point, boiling point, and vapor pressure.

43. B is correct.

Boyle's law (i.e., pressure-volume law) states that pressure and volume are inversely proportional:

$$(P_1V_1) = (P_2V_2)$$

or

$$P \times V = \text{constant}$$

If the volume of a gas increases, its pressure decreases proportionally.

44. E is correct.

Gases form homogeneous mixtures, regardless of the identities or relative proportions of the component gases. There is a relatively large distance between gas molecules (as opposed to solids or liquids where the molecules are much closer).

When pressure is applied to gas, its volume readily decreases, and thus gases are highly compressible.

There are no attractive forces between gas molecules, which is why molecules of gas can move about freely.

45. E is correct.

Solids have a definite shape and volume. For example, a granite block does not change its shape or volume regardless of the container in which it is placed.

Molecules in a solid are very tightly packed due to the strong intermolecular attractions, which prevent the molecules from moving around.

46. A is correct.

Standard temperature and pressure (STP) have a temperature of 273.15 K (0 °C, 32 °F) and a pressure of 10^5 Pa (100 kPa, 750.06 mmHg, 1 bar, 14.504 psi, 0.98692 atm).

The mm of Hg was defined as the pressure generated by a column of mercury one millimeter high. The pressure of mercury depends on temperature and gravity.

This variation in mmHg and torr is a difference in units of about 0.000015%.

In general,

1 torr = 1 mm of Hg = 0.0013158 atm.

750.06 mmHg = 0.98692 atm.

47. C is correct.

Intermolecular forces act between neighboring molecules. Examples include hydrogen bonding, dipole-dipole, dipole-induced dipole, and van der Waals (i.e., London dispersion) forces.

Stronger force results in a higher boiling point.

CH_3COOH is a carboxylic acid that can form two hydrogen bonds. Therefore, it has the highest boiling point.

Ethanoic acid with the two hydrogen bonds indicated on the structure

48. C is correct.

Molecules in solids have the most attraction to their neighbors, followed by liquids (significant motion between the individual molecules) and then gas.

A molecule in an ideal gas has no attraction to other gas molecules. For a gas experiencing low pressure, the particles are far enough apart for no attractive forces between the individual gas molecules.

49. C is correct.

Ideal gas law:

$$PV = nRT$$

where P is pressure, V is volume, n is the number of molecules, R is the ideal gas constant, and T is the temperature of the gas.

Set the initial and final P, V and T conditions equal:

$$(P_1V_1 / T_1) = (P_2V_2 / T_2)$$

Solve for the final volume of N_2:

$$(P_2V_2 / T_2) = (P_1V_1 / T_1)$$

$$V_2 = (T_2 P_1V_1) / (P_2T_1)$$

$$V_2 = [(295 \text{ K}) \times (750.06 \text{ mmHg}) \times (0.190 \text{ L } N_2)] / [(660 \text{ mmHg}) \times (298.15 \text{ K})]$$

$$V_2 = 0.214 \text{ L}$$

50. B is correct.

Volatility is the tendency of a substance to vaporize (phase change from liquid to vapor).

Volatility is directly related to a substance's vapor pressure. At a given temperature, a substance with higher vapor pressure vaporizes more readily than a substance with lower vapor pressure.

Molecules with weak intermolecular attraction can increase their kinetic energy by transferring less heat due to a smaller molecular mass. The increase in kinetic energy is required for individual molecules to move from the liquid to the gaseous phase.

51. B is correct.

Barometer and manometer are used to measure pressure.

Barometers are designed to measure atmospheric pressure, while a manometer can measure the pressure that is lower than atmospheric pressure.

A manometer has both ends of the tube open to the outside (while some may have one end closed), whereas a barometer is a type of closed-end manometer with one end of the glass tube closed and sealed with a vacuum.

The difference in mercury height on both necks indicates the capacity of a manometer.

$$820 \text{ mm} - 160 \text{ mm} = 660 \text{ mm}$$

Historically, the pressure unit of torr is set to equal 1 mmHg or the rise/dip of 1 mm of mercury in a manometer.

Because the manometer uses mercury, the height difference (660 mm) equals its measuring capacity in torr (660 torrs).

52. C is correct.

The conditions of the ideal gases are the same, so the number of moles (i.e., molecules) is equal.

At STP, the temperature is the same, so the kinetic energy of the molecules is the same.

However, the molar mass of oxygen and nitrogen are different. Therefore, the density is different.

53. A is correct.

Vapor pressure is the pressure exerted by a vapor in equilibrium with its condensed phases (i.e., solid or liquid) in a closed system at a given temperature.

The compound exists as a liquid if the external pressure > compound's vapor pressure.

A substance boils when vapor pressure = external pressure.

54. D is correct.

Gas molecules have a large amount of space between them; therefore, they can be pushed together, and gases are thus very compressible.

Molecules in solids and liquids are already close together; therefore, they cannot get significantly closer and are nearly incompressible.

55. C is correct.

Charles' law (i.e., the law of volumes) explains how, at constant pressure, gases behave when the temperature changes:

$$V \alpha T$$

or

$$V / T = constant$$

or

$$(V_1 / T_1) = (V_2 / T_2)$$

Volume and temperature are proportional. Therefore, an increase in one results in an increase in the other.

Gay-Lussac's law (i.e., pressure-temperature law) states that pressure is proportional to temperature:

$$P \alpha T$$

or

$$(P_1 / T_1) = (P_2 / T_2)$$

or

$$(P_1 T_2) = (P_2 T_1)$$

or

$$P / T = constant$$

If the pressure of a gas increases, the temperature increases.

Dalton's law (i.e., the law of partial pressures) states that *the pressure exerted by a mixture of gases is equal to the sum of the individual gas pressures*

Boyle's law (i.e., pressure-volume law) states that pressure and volume are inversely proportional:

$$(P_1V_1) = (P_2V_2)$$

or

$$P \times V = \text{constant}$$

If the volume of a gas increases, its pressure decreases proportionally.

Avogadro's law is an experimental gas law relating the volume of a gas to the amount of substance of gas present.

56. B is correct.

Sublimation is the direct change of state from a solid to a gas, skipping the intermediate liquid phase.

An example of a compound that undergoes sublimation is solid carbon dioxide (i.e., dry ice). CO_2 changes phases from solid to gas (i.e., bypasses the liquid phase) and is often used as a cooling agent.

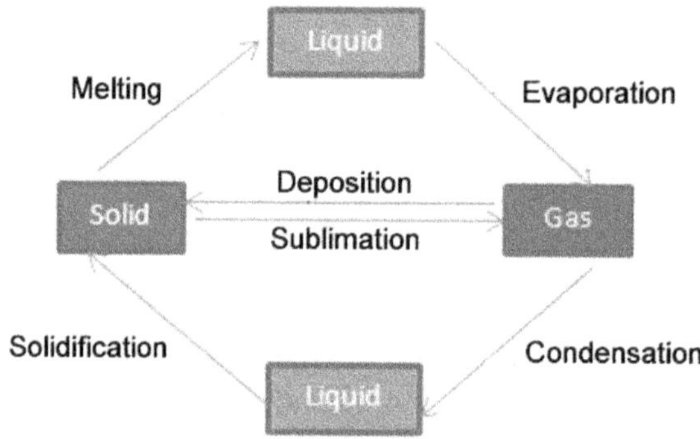

Interconversion of states of

57. B is correct.

Graham's law of effusion states that the rate of effusion (i.e., escaping through a small hole) of a gas is inversely proportional to the square root of the molar mass of its particles.

Rate 1 / Rate 2 = $\sqrt{\text{molar mass gas 2 / molar mass gas 1}}$

The diffusion rate is the inverse root of the molecular weights of the gases.

Therefore, the rate of effusion is:

O_2 / H_2 = $\sqrt{2 / 32}$

rate of diffusion = 1:4

58. B is correct.

To calculate the number of molecules, calculate the moles of gas using the ideal gas law:

$PV = nRT$

$n = PV / RT$

Because the gas constant R is in L·atm K^{-1} mol^{-1}, pressure must be converted into atm:

320 mmHg × (1 / 760 atm/mmHg) = 320 mmHg / 760 atm

(Leave it in this fraction form because the answer choices are in this format.)

Convert the temperature to Kelvin:

10 °C + 273 = 283 K

Substitute those values into the ideal gas equation:

$n = PV / RT$

n = (320 mmHg / 760 atm) × 6 L / (0.0821 L·atm K^{-1} mol^{-1} × 283 K)

This expression represents the number of gas moles present in the container.

Calculate the number of molecules:

number of molecules = moles × Avogadro's number

number of molecules = (320 mmHg / 760 atm) × 6 / (0.0821 × 283) × 6.02 × 10^{23}

number of molecules = (320 / 760)·(6)·(6 × 10^{23}) / (0.0821)·(283)

59. A is correct.

At STP (standard conditions for temperature and pressure), the pressure and temperature are the same regardless of the gas.

The ideal gas law:

$$PV = nRT$$

where P is pressure, V is volume, n is the number of molecules, R is the ideal gas constant, and T is the temperature of the gas.

Therefore, one molecule of each gas occupies the same volume at STP. Since CO_2 molecules have the largest mass, CO_2 gas has a greater mass in the same volume, and thus it has the greatest density.

60. C is correct.

Hydrogens, bonded directly to F, O or N, participate in hydrogen bonds.

The hydrogen is partially positive (i.e., delta plus: $\partial+$) due to the bond to these electronegative atoms. The lone pair of electrons on the F, O or N interacts with the partial positive ($\partial+$) hydrogen to form a hydrogen bond.

Practice Set 4: Questions 61–80

61. D is correct.

Calculate the moles of each gas:

moles = mass / molar mass

moles H_2 = 9.50 g / (2 × 1.01 g/mole)

moles H_2 = 4.70 moles

moles Ne = 14.0 g / (20.18 g/mole)

moles Ne = 0.694 moles

Calculate the mole fraction of H_2:

Mole fraction of H_2 = moles of H_2 / total moles in mixture

Mole fraction of H_2 = 4.70 moles / (4.70 moles + 0.694 moles)

Mole fraction of H_2 = 4.70 moles / (5.394 moles)

Mole fraction of H_2 = 0.87 moles

62. E is correct.

Vapor pressure is the pressure exerted by a vapor in equilibrium with
its condensed phases (i.e., solid or liquid) in a closed system at a given temperature.

Boiling occurs when the vapor pressure of a liquid equals atmospheric pressure.

Vapor pressure increases as the temperature increases.

Liquid A boils at a lower temperature than B because the vapor pressure of Liquid A is closer
to the atmospheric pressure.

63. B is correct.

Colligative properties of solutions depend on the ratio of the number of solute particles to the
number of solvent molecules in a solution and not on the type of chemical species present.

Colligative properties include: lowering of vapor pressure, the elevation of boiling point,
depression of freezing point, and increased osmotic pressure.

Freezing point (FP) depression:

$\Delta FP = -iKm$

where i = the number of particles produced when the solute dissociates, K = freezing point depression constant and m = molality (moles/kg solvent).

$$-iKm = -2 \text{ K}$$

$$-(1)\cdot(40)\cdot(x) = -2$$

$$x = -2 \, / -1(40)$$

$$x = 0.05 \text{ molal}$$

Assume that compound x does not dissociate.

$$0.05 \text{ mole compound } (x) \, / \text{ kg camphor} = 25 \text{ g} \, / \text{ kg camphor}$$

Therefore, if 0.05 mole = 25 g

$$1 \text{ mole} = 500 \text{ g}$$

64. E is correct.

Boiling occurs when the vapor pressure of a liquid equals atmospheric pressure.

Vapor pressure is the pressure exerted by a vapor in equilibrium with its condensed phases (i.e., solid or liquid) in a closed system at a given temperature.

Atmospheric pressure is the pressure exerted by the weight of air in the atmosphere.

Vapor pressure is inversely correlated with the strength of the intermolecular force.

The molecules are more likely to stick together in liquid form with stronger intermolecular forces.

Few participate in the liquid-vapor equilibrium; therefore, the molecule would boil at a higher temperature.

65. D is correct.

Van der Waals equation describes factors that must be accounted for when the ideal gas law calculates values for nonideal gases.

The terms that affect the pressure and volume of the ideal gas law are intermolecular forces and the volume of nonideal gas molecules.

66. E is correct.

Ideal gas law:

$$PV = nRT$$

where P is pressure, V is volume, n is the number of molecules, R is the ideal gas constant, and T is the temperature of the gas.

Units of R can be calculated by rearranging the expression:

$$R = PV/nT$$

$$R = atm \cdot L/mol \cdot K$$

67. A is correct.

Ideal gas law:

$$PV = nRT$$

where P is pressure, V is volume, n is the number of molecules, R is the ideal gas constant, and T is the temperature of the gas.

Set the initial and final P/V/T conditions equal:

$$(P_1 V_1 / T_1) = (P_2 V_2 / T_2)$$

STP condition is the temperature of 0 °C (273 K) and pressure of 1 atm.

Solve for the final temperature:

$$(P_2 V_2 / T_2) = (P_1 V_1 / T_1)$$

$$T_2 = (P_2 V_2 T_1) / (P_1 V_1)$$

$$T_2 = [(0.80 \text{ atm}) \times (0.155 \text{ L}) \times (273 \text{ K})] / [(1.00 \text{ atm}) \times (0.120 \text{ L})]$$

$$T_2 = 282.1 \text{ K}$$

Then convert temperature units to degrees Celsius:

$$T_2 = (282.1 \text{ K} - 273 \text{ K}) = 9.1 \text{ °C}$$

68. C is correct.

Atmospheric pressure is the pressure exerted by the weight of air in the atmosphere. Boiling occurs when the vapor pressure of the liquid is higher than the atmospheric pressure.

At standard atmospheric pressure and 22 °C, water vapor pressure is less than the atmospheric pressure, and it does not boil. However, when a vacuum pump is used, the atmospheric pressure is reduced until it has a lower vapor pressure than water, allowing water to boil at a much lower temperature.

69. E is correct.

The kinetic theory of gases describes a gas as a large number of small particles in constant rapid motion. These particles collide with each other and with the walls of the container.

Their average kinetic energy depends only on the absolute temperature of the system.

$$\text{temperature} = \tfrac{1}{2}mv^2$$

At high temperatures, the particles are moving greater velocity, and at a temperature of absolute zero (i.e., 0 K), there is no movement of gas particles.

Therefore, as temperature decreases, kinetic energy decreases, and so does the velocity of the gas molecules.

70. A is correct.

Dalton's law (i.e., the law of partial pressures) states that *the pressure exerted by a mixture of gases is equal to the sum of the individual gas pressures.*

The partial pressure of molecules in a mixture is proportional to their molar ratios.

Use the coefficients of the reaction to determine the molar ratio.

Based on that information, calculate the partial pressure of O_2:

(coefficient O_2) / (sum of coefficients in the mixture) × total pressure

$[1 / (2 + 1)] \times 1{,}250$ torr

417 torr = partial pressure of O_2

71. D is correct.

$$2\,Na\,(s) + Cl_2\,(g) \rightarrow 2\,NaCl\,(s)$$

In its elemental form, Na exists as a solid.

In its elemental form, chlorine exists as a gas.

72. B is correct.

Charles' law (i.e., the law of volumes) explains how, at constant pressure, gases behave when the temperature changes:

$$V \propto T$$

or

$$V / T = \text{constant}$$

or

$$(V_1 / T_1) = (V_2 / T_2)$$

Volume and temperature are proportional.

Therefore, an increase in one term increases the other.

73. E is correct.

Gay-Lussac's law (i.e., pressure-temperature law) states that pressure is proportional to temperature:

$$P \propto T$$

or $\quad (P_1 / T_1) = (P_2 / T_2)$

or $\quad (P_1 T_2) = (P_2 T_1)$

or $\quad P / T = \text{constant}$

If the pressure of a gas increases, the temperature is increased proportionally.

Quadrupling the temperature increases the pressure fourfold.

Boyle's law (i.e., pressure-volume law) states that pressure and volume are inversely proportional:

$$(P_1 V_1) = (P_2 V_2)$$

or $\quad P \times V = \text{constant}$

Reducing the volume by half increases pressure twofold.

Therefore, the increase in pressure would be by a factor of $4 \times 2 = 8$.

74. C is correct.

Scientists found that the relationships between pressure, temperature, and volume of a sample of gas hold for gases, and the gas laws were developed.

Boyle's law (i.e., pressure-volume law) states that pressure and volume are inversely proportional:

$$(P_1 V_1) = (P_2 V_2)$$

or

$$P \times V = \text{constant}$$

If the volume of a gas increases, its pressure decreases proportionally.

Charles' law (i.e., the law of volumes) explains how, at constant pressure, gases behave when the temperature changes:

$$(V_1 / T_1) = (V_2 / T_2)$$

Gay-Lussac's law (i.e., pressure-temperature law) states that pressure is proportional to temperature:

$$P \, \alpha \, T$$

or

$$(P_1 / T_1) = (P_2 / T_2) \text{ or } (P_1 T_2) = (P_2 T_1)$$

or

$$P / T = \text{constant}$$

If the pressure of a gas increases, the temperature increases.

75. A is correct.

Sublimation is the direct change of state from a solid to a gas, skipping the intermediate liquid phase.

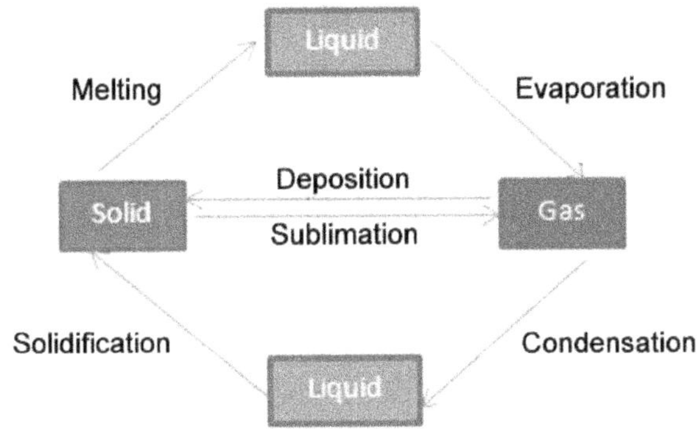

Interconversion of states of matter

An example of a compound that undergoes sublimation is solid carbon dioxide (i.e., dry ice). CO_2 changes phases from solid to gas (i.e., bypasses the liquid phase) and is often used as a cooling agent.

76. B is correct.

Graham's law of effusion states that the rate of effusion (i.e., escaping through a small hole) of a gas is inversely proportional to the square root of the molar mass of its particles.

$$\text{Rate 1 / Rate 2} = \sqrt{(\text{molar mass gas 2 / molar mass gas 1})}$$

Set the rate of effusion of krypton over the rate of effusion of methane:

$$\text{Rate}_{Kr} / \text{Rate}_{CH4} = \sqrt{[(M_{Kr}) / (M_{CH4})]}$$

Solve for the ratio of effusion rates:

$$\text{Rate}_{Kr} / \text{Rate}_{CH4} = \sqrt{[(83.798 \text{ g/mol}) / (16.04 \text{ g/mol})]}$$

$$\text{Rate}_{Kr} / \text{Rate}_{CH4} = 2.29$$

Larger gas molecules must effuse at a slower rate; the effusion rate of Kr gas molecules must be slower than methane molecules:

$$\text{Rate}_{Kr} = [\text{Rate}_{CH4} / (2.29)]$$

$$\text{Rate}_{Kr} = [(631 \text{ m/s}) / (2.29)]$$

$$\text{Rate}_{Kr} = 276 \text{ m/s}$$

77. C is correct.

Observe information about gas C. The temperature, pressure, and volume of the gas are indicated.

Substitute these values into the ideal gas law equation to determine the number of moles:

$$PV = nRT$$

$$n = PV / RT$$

Convert units to the units indicated in the gas constant:

volume = 668.5 mL × (1 L / 1,000 mL)

volume = 0. 669 L

pressure = 745.5 torr × (1 atm / 760 torr)

pressure = 0.981 atm

temperature = 32.0 °C + 273.15 K

temperature = 305.2 K

$$n = PV / RT$$

$$n = (0.981 \text{ atm} × 0.669 \text{ L}) / (0.0821 \text{ L·atm K}^{-1} \text{mol}^{-1} × 305.2 \text{ K})$$

n = 0.026 mole

The *law of mass conservation* states that the mass of products = the mass of reactants.

Based on this law, the mass of gas C can be determined:

mass product = mass reactant

mass A = mass B + mass C

5.2 g = 3.8 g + mass C

mass C = 1.4 g

Calculate the molar mass of C:

molar mass = 1.4 g / 0.026 mole

molar mass = 53.9 g / mole

78. E is correct.

Liquids take the shape of their container (i.e., they have an indefinite shape). This can be visualized by considering a liter of water poured into a cylindrical bucket or a square box. In both cases, it takes up the shape of the container.

Liquids have a definite volume. The volume of water in the given example is 1 L, regardless of its container.

79. D is correct.

Intermolecular forces act between neighboring molecules. Examples include hydrogen bonding, dipole-dipole, dipole-induced dipole, and van der Waals (i.e., London dispersion) forces.

Hydrogens, bonded directly to F, O or N, participate in hydrogen bonds.

The hydrogen is partially positive (i.e., delta plus: $\partial+$) due to the bond to these electronegative atoms.

The lone pair of electrons on the F, O or N interacts with the partial positive ($\partial+$) hydrogen to form a hydrogen bond.

80. D is correct.

When the balloon is placed in a freezer, the temperature of the helium in the balloon decreases because the surroundings are colder than the balloon, and heat is transferred from the balloon to the surrounding air until a thermal equilibrium is reached.

When the temperature of a gas decreases, the molecules move more slowly and become closer together, causing the volume of the balloon to decrease.

From the ideal gas law,

$$PV = nRT$$

If temperature decreases, the volume must decrease (if pressure remains constant).

Notes for active learning

Practice Set 5: Questions 81–100

81. A is correct.

Boyle's law (i.e., pressure-volume law) states that pressure and volume are inversely proportional:

$$(P_1 V_1) = (P_2 V_2)$$

or

$$P \times V = \text{constant}$$

If the pressure of a gas increases, its volume decreases proportionally.

Doubling the pressure reduces the volume by half.

82. C is correct.

Vapor pressure is the pressure exerted by a vapor in equilibrium with its condensed phases (i.e., solid or liquid) in a closed system at a given temperature.

The pressure of a gas is the force that it exerts on the walls of its container. This is essentially the frequency and energy with which the gas molecules collide with the container's walls.

Partial pressure refers to the individual pressure of a gas that is in a mixture of gases.

Vapor pressure is the pressure of a vapor in thermodynamic equilibrium with its solid or liquid phases.

83. B is correct.

Colligative properties of solutions depend on the ratio of the number of solute particles to the number of solvent molecules in a solution and not on the type of chemical species present.

Colligative properties include: lowering of vapor pressure, the elevation of boiling point, depression of freezing point, and increased osmotic pressure.

Freezing point (FP) depression:

$$\Delta FP = -iKm$$

where i = the number of particles produced when the solute dissociates, K = freezing point depression constant, and m = molality (moles/kg solvent).

Given in the question stem:

$$\Delta FP = -10 \text{ K}$$

$$-iKm = -10 \text{ K}$$

Toluene is the solvent and does not dissociate; the K value is not used.

FP depression from the addition of benzene:

$$i = 10 \text{ K} / Km$$

$$i = 1 - (1)\cdot(5.0)\cdot(x)$$

K value of solvent:

$$x = 2 \text{ molal}$$

2 moles toluene/kg benzene $= x$ moles / 0.1 kg benzene

x moles / 0.1 kg benzene $= 2$ moles toluene/kg benzene

x moles $= 2$ moles toluene/kg benzene $\times$ 0.1 kg benzene

toluene $= 0.2$ mole

84. E is correct.

Considering the ideal gas law, there is no difference between He and Ne since pressure, volume, temperature, and the number of moles are known quantities.

The identity of the gas would only be relevant to convert from moles to mass or vice-versa.

85. D is correct.

Gas molecules transfer kinetic energy between each other.

86. D is correct.

Assume a is the equivalent temperature (i.e., has the same value in °C and °F).

Set up a formula with the equivalent temperature on one side and the conversion factor to the other unit on the other side.

For example, the left side is °C, while the right side is the conversion to °F:

$$a = [(9/5) \times a] + 32$$

$$a - (9/5)a = 32$$

$$(-4/5)a = 32$$

$$-4a = 160$$

$$a = -40$$

At –40, the temperature is the same in °C and °F.

87. D is correct.

At low elevations (i.e., sea level), the boiling point of water is 100 °C.

Atmospheric pressure is the pressure exerted by the weight of air in the atmosphere.

The boiling point decreases with increasing altitude due to the reduced atmospheric pressure above the water at high altitudes.

88. A is correct.

Hydrocarbons are nonpolar molecules, which means that the dominant intermolecular force is London dispersion. This force increases as the number of atoms in each molecule increases.

Stronger force results in a higher boiling point.

CH_4 has the least atoms and therefore has the lowest boiling point.

89. C is correct.

The melting point of a substance depends on the pressure that it is at.

The triple point (point D on the graph) is the pressure and temperature at which a substance exists as a solid, liquid, and gas.

The critical point (point E on the graph) is the endpoint of the phase equilibrium curve where the liquid and its vapor become indistinguishable.

Generally, melting points are given at standard pressure. If the substance is denser in the solid than in the liquid state (which is true of most substances), the melting point increases when pressure is increased.

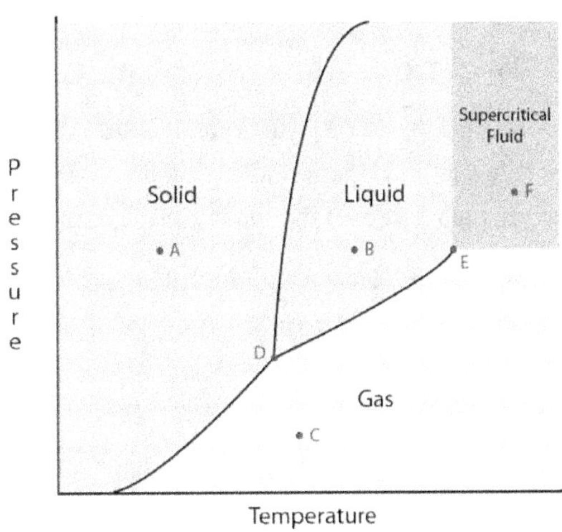

Phase diagram of pressure vs. temperature

This is observed with the phase diagram for CO_2. However, with certain substances, including water, the solid is less dense than the liquid state (e.g., ice is less dense than water), so the reverse is true, and the melting point decreases when pressure increases.

90. E is correct.

Dalton's law (i.e., the law of partial pressures) states that *the pressure exerted by a mixture of gases is equal to the sum of the individual gas pressures.*

The partial pressure of molecules in a mixture is proportional to their molar ratios.

Moles of CH_4: 32 g × 16 g/mol = 2 moles of CH_4

Moles of NH_3: 12.75 g × 17 g/mole = 0.75 mole of NH_3

Mole fraction of NH_3 gas:

mole fraction of NH_3 = (moles of NH_3) / (total moles in mixture)

mole fraction of NH_3 = (0.75 mole) / (2.75 moles)

mole fraction of NH_3 = 0.273

Partial pressure of NH_3 gas:

mole fraction of NH_3 × total pressure

(0.273)·(2.4 atm) = 0.66 atm

91. B is correct.

Liquids take the shape of the container they are in (i.e., they have an indefinite shape). This can be visualized by considering a liter of water poured into a cylindrical bucket or a square box. In both cases, it takes up the shape of the container.

However, liquids have a definite volume. The volume of water in the example is 1 L, regardless of its container.

92. E is correct.

Charles' law (i.e., the law of volumes) explains how, at constant pressure, gases behave when the temperature changes:

$(V_1 / T_1) = (V_2 / T_2)$

Solve for the final temperature:

$T_2 = (V_1 / T_1) / V_2$

$T_2 = (0.050 \text{ L}) × [(420 \text{ K}) / (0.100 \text{ L})]$

$T_2 = 210 \text{ K}$

93. A is correct.

Gay-Lussac's law (i.e., pressure-temperature law) states that pressure is proportional to temperature:

$$P \, \alpha \, T$$

or

$$(P_1 \, / \, T_1) = (P_2 \, / \, T_2)$$

or

$$(P_1 T_2) = (P_2 T_1)$$

or

$$P \, / \, T = \text{constant}$$

If the pressure of a gas increases, the temperature increases.

Dalton's law (i.e., the law of partial pressures) states that *the pressure exerted by a mixture of gases is equal to the sum of the individual gas pressures*.

Charles' law (i.e., the law of volumes) explains how, at constant pressure, gases behave when the temperature changes:

$$V \, \alpha \, T \text{ or } V \, / \, T = \text{constant}$$

Volume and temperature are proportional.

Therefore, an increase in one term increases the other.

Boyle's law (i.e., pressure-volume law) states that pressure and volume are inversely proportional:

$$(P_1 V_1) = (P_2 V_2) \text{ or } P \times V = \text{constant}$$

If the volume of a gas increases, its pressure decreases proportionally.

94. E is correct.

Gay-Lussac's law (i.e., pressure-temperature law) states that pressure is proportional to temperature:

$$P \, \alpha \, T$$

or

$$(P_1 \, / \, T_1) = (P_2 \, / \, T_2)$$

or

$$(P_1 T_2) = (P_2 T_1)$$

or

$$P / T = \text{constant}$$

If the pressure of a gas increases, the temperature increases.

The kinetic theory of gases describes a gas as a large number of small particles in constant rapid motion. These particles collide with each other and with the walls of the container.

Their average kinetic energy depends only on the absolute temperature of the system.

$$\text{temperature} = \tfrac{1}{2}mv^2$$

At high temperatures, the particles are moving with greater velocity.

95. C is correct.

Vaporization refers to the change of state from a liquid to a gas.

There are two types of vaporization: boiling and evaporation, which are differentiated based on the temperature at which they occur.

Evaporation occurs at a temperature below the boiling point, while boiling occurs at or above the boiling point.

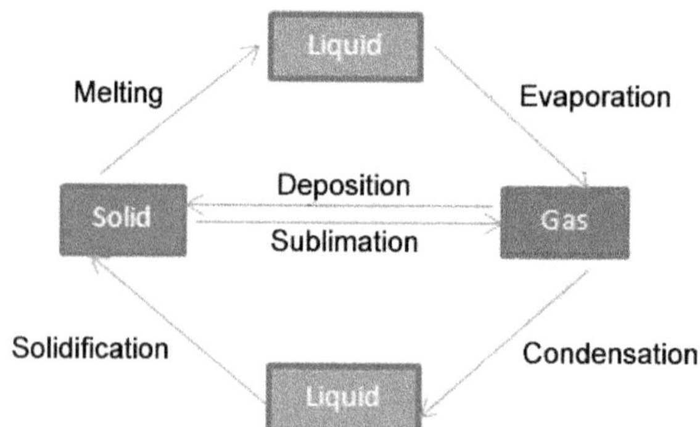

Interconversion of states of matter

96. C is correct.

Intermolecular forces act between neighboring molecules. Examples include hydrogen bonding, dipole-dipole, dipole-induced dipole, and van der Waals (i.e., London dispersion) forces.

A dipole-dipole attraction involves asymmetric, polar molecules (based on differences in electronegativity between atoms) that create a dipole moment (i.e., a net vector indicating the force).

Dimethyl ether with an indicated bond angle

97. E is correct.

Ideal gas law:

$$PV = nRT$$

where P is pressure, V is volume, n is the number of molecules, R is the ideal gas constant, and T is the temperature of the gas.

R and n are constant.

Therefore, decreasing the volume, increasing temperature, or increasing the number of molecules increases the pressure of a gas.

98. E is correct.

Avogadro's law is an experimental gas law relating the volume of a gas to the amount of substance of gas present. It states that "equal volumes of all gases, at the same temperature and pressure, have the same number of molecules."

For a given mass of an ideal gas, the gas's volume and amount (moles) are directly proportional if the temperature and pressure are constant.

$$V \, \alpha \, n$$

or

$$V \, / \, n = k$$

where V = volume of the gas, n is the amount of substance of the gas (measured in moles).

k is a constant equal to RT / P,

where R is the universal gas constant, T is the Kelvin temperature, and P is the pressure.

As temperature and pressure are constant, RT / P is constant and represented as k (derived from the ideal gas law).

At the same temperature and pressure, equal volumes of gases have the same number of molecules.

Comparing the same substance under two different sets of conditions:

$$V_1 / n_1 = V_2 / n_2$$

As the number of moles of gas increases, the volume of the gas increases in proportion.

Similarly, if the number of moles of gas is decreased, then the volume decreases.

Thus, the number of molecules or atoms in a specific volume of an ideal gas is independent of their size or the molar mass of the gas.

99. B is correct.

The ideal gas law: $PV = nRT$

Using proportions, if the temperature is increased by a factor of 1.5 and the pressure is increased by a factor of 1.5 (constant number of moles), volume remains the same.

100. C is correct.

Since both He and Ne are at the same temperature, pressure, and volume, they have equal moles of gas.

Even though there are equal moles of each gas of He and Ne, use the molar masses given in the periodic table and notice that the samples contain different masses.

Notes for active learning

Notes for active learning

APPENDIX

Periodic Table of the Elements

Key: atomic number / **Symbol** / name / abridged standard atomic weight

1	2	3	4	5	6	7	8	9	10	11	12	13	14	15	16	17	18
1 **H** hydrogen 1.0080 ±0.0002																	2 **He** helium 4.0026 ±0.0001
3 **Li** lithium 6.94 ±0.06	4 **Be** beryllium 9.0122 ±0.0001											5 **B** boron 10.81 ±0.02	6 **C** carbon 12.011 ±0.002	7 **N** nitrogen 14.007 ±0.001	8 **O** oxygen 15.999 ±0.001	9 **F** fluorine 18.998 ±0.001	10 **Ne** neon 20.180 ±0.001
11 **Na** sodium 22.990 ±0.001	12 **Mg** magnesium 24.305 ±0.002											13 **Al** aluminium 26.982 ±0.001	14 **Si** silicon 28.085 ±0.001	15 **P** phosphorus 30.974 ±0.001	16 **S** sulfur 32.06 ±0.02	17 **Cl** chlorine 35.45 ±0.01	18 **Ar** argon 39.95 ±0.16
19 **K** potassium 39.098 ±0.001	20 **Ca** calcium 40.078 ±0.004	21 **Sc** scandium 44.956 ±0.001	22 **Ti** titanium 47.867 ±0.001	23 **V** vanadium 50.942 ±0.001	24 **Cr** chromium 51.996 ±0.001	25 **Mn** manganese 54.938 ±0.001	26 **Fe** iron 55.845 ±0.002	27 **Co** cobalt 58.933 ±0.001	28 **Ni** nickel 58.693 ±0.001	29 **Cu** copper 63.546 ±0.003	30 **Zn** zinc 65.38 ±0.02	31 **Ga** gallium 69.723 ±0.001	32 **Ge** germanium 72.630 ±0.008	33 **As** arsenic 74.922 ±0.001	34 **Se** selenium 78.971 ±0.008	35 **Br** bromine 79.904 ±0.003	36 **Kr** krypton 83.798 ±0.002
37 **Rb** rubidium 85.468 ±0.001	38 **Sr** strontium 87.62 ±0.01	39 **Y** yttrium 88.906 ±0.001	40 **Zr** zirconium 91.224 ±0.002	41 **Nb** niobium 92.906 ±0.001	42 **Mo** molybdenum 95.95 ±0.01	43 **Tc** technetium [97]	44 **Ru** ruthenium 101.07 ±0.02	45 **Rh** rhodium 102.91 ±0.01	46 **Pd** palladium 106.42 ±0.01	47 **Ag** silver 107.87 ±0.01	48 **Cd** cadmium 112.41 ±0.01	49 **In** indium 114.82 ±0.01	50 **Sn** tin 118.71 ±0.01	51 **Sb** antimony 121.76 ±0.01	52 **Te** tellurium 127.60 ±0.03	53 **I** iodine 126.90 ±0.01	54 **Xe** xenon 131.29 ±0.01
55 **Cs** caesium 132.91 ±0.01	56 **Ba** barium 137.33 ±0.01	57–71 lanthanoids	72 **Hf** hafnium 178.49 ±0.01	73 **Ta** tantalum 180.95 ±0.01	74 **W** tungsten 183.84 ±0.01	75 **Re** rhenium 186.21 ±0.01	76 **Os** osmium 190.23 ±0.03	77 **Ir** iridium 192.22 ±0.01	78 **Pt** platinum 195.08 ±0.02	79 **Au** gold 196.97 ±0.01	80 **Hg** mercury 200.59 ±0.01	81 **Tl** thallium 204.38 ±0.01	82 **Pb** lead 207.2 ±1.1	83 **Bi** bismuth 208.98 ±0.01	84 **Po** polonium [209]	85 **At** astatine [210]	86 **Rn** radon [222]
87 **Fr** francium [223]	88 **Ra** radium [226]	89–103 actinoids	104 **Rf** rutherfordium [267]	105 **Db** dubnium [268]	106 **Sg** seaborgium [269]	107 **Bh** bohrium [270]	108 **Hs** hassium [269]	109 **Mt** meitnerium [277]	110 **Ds** darmstadtium [281]	111 **Rg** roentgenium [282]	112 **Cn** copernicium [285]	113 **Nh** nihonium [286]	114 **Fl** flerovium [290]	115 **Mc** moscovium [290]	116 **Lv** livermorium [293]	117 **Ts** tennessine [294]	118 **Og** oganesson [294]

57 **La** lanthanum 138.91 ±0.01	58 **Ce** cerium 140.12 ±0.01	59 **Pr** praseodymium 140.91 ±0.01	60 **Nd** neodymium 144.24 ±0.01	61 **Pm** promethium [145]	62 **Sm** samarium 150.36 ±0.02	63 **Eu** europium 151.96 ±0.01	64 **Gd** gadolinium 157.25 ±0.03	65 **Tb** terbium 158.93 ±0.01	66 **Dy** dysprosium 162.50 ±0.01	67 **Ho** holmium 164.93 ±0.01	68 **Er** erbium 167.26 ±0.01	69 **Tm** thulium 168.93 ±0.01	70 **Yb** ytterbium 173.05 ±0.02	71 **Lu** lutetium 174.97 ±0.01
89 **Ac** actinium [227]	90 **Th** thorium 232.04 ±0.01	91 **Pa** protactinium 231.04 ±0.01	92 **U** uranium 238.03 ±0.01	93 **Np** neptunium [237]	94 **Pu** plutonium [244]	95 **Am** americium [243]	96 **Cm** curium [247]	97 **Bk** berkelium [247]	98 **Cf** californium [251]	99 **Es** einsteinium [252]	100 **Fm** fermium [257]	101 **Md** mendelevium [258]	102 **No** nobelium [259]	103 **Lr** lawrencium [262]

International Union of Pure and Applied Chemistry (IUPAC), 4 May 2022

Notes for active learning

Common Chemistry Equations

Throughout the test the following symbols have the definitions specified unless otherwise noted.

L, mL	= liter(s), milliliter(s)		mm Hg	= millimeters of mercury
g	= gram(s)		J, kJ	= joule(s), kilojoule(s)
nm	= nanometer(s)		V	= volt(s)
atm	= atmosphere(s)		mol	= mole(s)

ATOMIC STRUCTURE

$$E = h\nu$$

$$c = \lambda\nu$$

E = energy
ν = frequency
λ = wavelength

Planck's constant, $h = 6.626 \times 10^{-34}$ J s

Speed of light, $c = 2.998 \times 10^{8}$ m s^{-1}

Avogadro's number $= 6.022 \times 10^{23}$ mol^{-1}

Electron charge, $e = -1.602 \times 10^{-19}$ coulomb

EQUILIBRIUM

$$K_c = \frac{[C]^c[D]^d}{[A]^a[B]^b}, \text{ where } a\,A + b\,B \rightleftarrows c\,C + d\,D$$

$$K_p = \frac{(P_C)^c(P_D)^d}{(P_A)^a(P_B)^b}$$

$$K_a = \frac{[H^+][A^-]}{[HA]}$$

$$K_b = \frac{[OH^-][HB^+]}{[B]}$$

$$K_w = [H^+][OH^-] = 1.0 \times 10^{-14} \text{ at } 25°C$$
$$= K_a \times K_b$$

$$pH = -\log[H^+], \quad pOH = -\log[OH^-]$$

$$14 = pH + pOH$$

$$pH = pK_a + \log\frac{[A^-]}{[HA]}$$

$$pK_a = -\log K_a, \quad pK_b = -\log K_b$$

Equilibrium Constants

K_c (molar concentrations)
K_p (gas pressures)
K_a (weak acid)
K_b (weak base)
K_w (water)

KINETICS

$$\ln[A]_t - \ln[A]_0 = -kt$$

$$\frac{1}{[A]_t} - \frac{1}{[A]_0} = kt$$

$$t_{1/2} = \frac{0.693}{k}$$

k = rate constant
t = time
$t_{1/2}$ = half-life

GASES, LIQUIDS, AND SOLUTIONS

$$PV = nRT$$

$$P_A = P_{total} \times X_A, \text{ where } X_A = \frac{\text{moles A}}{\text{total moles}}$$

$$P_{total} = P_A + P_B + P_C + \ldots$$

$$n = \frac{m}{M}$$

$$K = {}^\circ C + 273$$

$$D = \frac{m}{V}$$

$$KE \text{ per molecule} = \frac{1}{2}mv^2$$

Molarity, M = moles of solute per liter of solution

$$A = abc$$

P = pressure
V = volume
T = temperature
n = number of moles
m = mass
M = molar mass
D = density
KE = kinetic energy
v = velocity
A = absorbance
a = molar absorptivity
b = path length
c = concentration

Gas constant, R = 8.314 J mol^{-1} K^{-1}
$\qquad$ = 0.08206 L atm mol^{-1} K^{-1}
$\qquad$ = 62.36 L torr mol^{-1} K^{-1}
1 atm = 760 mm Hg
$\qquad$ = 760 torr
STP = $0.00\,^\circ$C and 10^5 Pa

THERMOCHEMISTRY/ ELECTROCHEMISTRY

$$q = mc\Delta T$$

$$\Delta S^\circ = \sum S^\circ \text{ products} - \sum S^\circ \text{ reactants}$$

$$\Delta H^\circ = \sum \Delta H_f^\circ \text{ products} - \sum \Delta H_f^\circ \text{ reactants}$$

$$\Delta G^\circ = \sum \Delta G_f^\circ \text{ products} - \sum \Delta G_f^\circ \text{ reactants}$$

$$\Delta G^\circ = \Delta H^\circ - T\Delta S^\circ$$

$$= -RT \ln K$$

$$= -n F E^\circ$$

$$I = \frac{q}{t}$$

q = heat
m = mass
c = specific heat capacity
T = temperature
S° = standard entropy
H° = standard enthalpy
G° = standard free energy
n = number of moles
E° = standard reduction potential
I = current (amperes)
q = charge (coulombs)
t = time (seconds)

Faraday's constant, F = 96,485 coulombs per mole of electrons

$$1 \text{ volt} = \frac{1 \text{ joule}}{1 \text{ coulomb}}$$

Annotated Glossary of Chemistry Terms

A

Absolute entropy (of a substance) – the increase in the entropy of a substance as it goes from a perfectly ordered crystalline form at 0 K (where its entropy is zero) to the temperature in question.

Absolute zero – the zero point on the absolute temperature scale; -273.15 °C or 0 K; theoretically, the temperature at which molecular motion ceases (i.e., the system does not emit or absorb energy, atoms at rest).

Absorption spectrum – spectrum associated with the absorption of electromagnetic radiation by atoms (or other species), resulting from transitions from lower to higher energy states.

Accuracy – the degree to which a value is close to the actual value; see *precision*.

Acid – a substance that produces H^+ (*aq*) ions in an aqueous solution and gives a pH of less than 7.0; strong acids ionize entirely or almost entirely in dilute aqueous solution; weak acids ionize only slightly. It turns litmus red.

Acid dissociation constant – an equilibrium constant for dissociating a weak acid.

Acid rain – rainwater with a pH of less than 5.7; caused by the gases NO_2 (vehicle exhaust fumes) and SO_2 (from burning fossil fuels) dissolving in the rain. It kills fish, wildlife, and trees and destroys buildings and lakes.

Acidic salt – contains an ionizable hydrogen atom; does not necessarily produce acidic solutions.

Actinides – the fifteen chemical elements that are between actinium (89) and lawrencium (103).

Activated complex – a structure formed by a collision between molecules in which new bonds form.

Activation energy – the amount of energy that reactants must absorb in their ground states to reach the transition state needed for a reaction to occur.

Active metal – a metal with low ionization energy that loses electrons readily to form cations.

Activity (of a component of ideal mixture) – a dimensionless quantity whose magnitude is equal to the molar concentration in an ideal solution; equal to partial pressure in an ideal gas mixture; 1 for pure solids or liquids.

Activity series – a listing of metals (and hydrogen) in order of decreasing activity.

Actual yield – the amount of a specified pure product obtained from a given reaction; see *theoretical yield*.

Addition reaction – a reaction in which two atoms or groups of atoms are added to a molecule, one on each side of a double or triple bond.

Adhesive forces – forces of attraction between a liquid and another surface.

Adsorption – the adhesion of a species onto the surfaces of particles.

Aeration – the mixing of air into a liquid or a solid.

Alcohol – a hydrocarbon derivative containing a ~OH group attached to a carbon atom, not in an aromatic ring.

Alkali metals – the elements of Group IA on the periodic table (e.g., Na, K, Rb).

Alkaline battery – a dry cell in which the electrolyte contains KOH.

Alkaline earth metals – group IIA metals on the periodic table; see *earth metals*.

Allomer – a substance that has a different composition from another but the same crystalline structure.

Allotropes – elements with different structures (therefore different forms), such as carbon (e.g., diamond, graphite, and fullerene).

Allotropic modifications (allotropes) – different forms of the same element in the same physical state.

Alloy – a mixture of metals. For example, bronze is an alloy formed from copper and tin.

Alloying – mixing metals with other substances (usually other metals) to modify their properties.

Alpha (α) particle – a helium nucleus; helium ion with 2+ charge; an assembly of two protons and two neutrons.

Amorphous solid – a non-crystalline solid with no well-defined ordered structure.

Ampere – unit of electrical current; one ampere equals one coulomb per second.

Amphiprotism – the ability of a substance to exhibit amphiprotic by accepting donated protons.

Amphoterism – the ability to react with both acids and bases; to act as either an acid or a base.

Amplitude – the maximum distance that medium particles carrying the wave move from their rest position.

Anion – a negative ion; an atom or group of atoms that has gained one or more electrons.

Anode – in a cathode ray tube, the positive electrode (electrode at which oxidation occurs); the positive side of a dry cell battery or a cell.

Antibonding orbital – a molecular orbital higher in energy than any of the atomic orbitals from which it is derived; lends instability to a molecule or ion when populated with electrons; denoted with star (*) superscript.

Artificial transmutation – an artificially induced nuclear reaction caused by the bombardment of a nucleus with subatomic particles or small nuclei.

Associated ions – short-lived species formed by the collision of dissolved ions of opposite charges.

Atmosphere – a unit of pressure; the pressure supports a column of mercury 760 mm high at 0 °C.

Atom – a chemical element in its smallest form; it comprises neutrons and protons within the nucleus and electrons circling the nucleus; it is the smallest part of an element that can exist.

Atomic mass unit (amu) – one-twelfth of the mass of an atom of the carbon-12 isotope; used for stating atomic and formula weights; known as a dalton.

Atomic number – represents an element corresponding with the number of protons within the nucleus; the number of protons in the nucleus of the atom.

Atomic orbital (*AO*) – a region or volume in space where the probability of finding electrons is highest.

Atomic radius – radius of an atom.

Atomic weight – weighted average of the masses of the constituent isotopes of an element; the relative masses of atoms of different elements.

Aufbau (or *building up*) principle – describes the order in which electrons fill orbitals in atoms.

Autoionization – an ionization reaction between identical molecules.

Avogadro's Law – equal volumes of gases contain the same number of molecules at the same temperature and pressure.

Avogadro's number (N_A) – the number (6.022×10^{23}) of atoms, molecules, or particles found in precisely 1 mole of a substance.

B

Background radiation – extraneous to an experiment; usually the low-level natural radiation from cosmic rays and trace radioactive substances present in the environment.

Band – a series of very closely spaced nearly continuous molecular orbitals that belong to the crystal as a whole.

Band of stability – band containing nonradioactive nuclides in a plot of neutrons *vs.* their atomic number.

Band theory of metals – the theory that accounts for the bonding and properties of metallic solids.

Barometer – a device used to measure the pressure in the atmosphere.

Base – a substance that produces ^-OH (*aq*) ions in an aqueous solution; accepts a proton and has a high pH; strongly soluble bases are soluble in water and are entirely dissociated; weak bases ionize only slightly; a typical example of a base is sodium hydroxide (NaOH). It turns litmus blue.

Basic anhydride – the oxide of a metal that reacts with water to form a base.

Basic salt – a salt containing an ionizable OH group.

Beta (β) particle – an electron emitted from the nucleus when a neutron decays to a proton and an electron.

Binary acid – a binary compound in which H is bonded to one or more electronegative nonmetals.

Binary compound – consists of two elements; it may be ionic or covalent.

Binding energy (nuclear binding energy) – the energy equivalent ($E = mc^2$) of the mass deficiency of an atom (where E is the energy in joules, m is the mass in kilograms, and c is the speed of light in m/s^2).

Boiling – the phase transition of a liquid vaporizing.

Boiling point – the temperature at which the vapor pressure of a liquid is equal to the applied pressure; the *condensation point*.

Boiling point elevation – the increase in the boiling point of a solvent caused by the dissolution of a nonvolatile solute.

Bomb calorimeter – a device used to measure the heat transfer between a system and its surroundings at constant volume.

Bond – the attraction and repulsion between atoms and molecules is a cornerstone of chemistry.

Bond energy – the amount of energy necessary to break one mole of bonds in a substance, dissociating the substance in its gaseous state into atoms of its elements in the gaseous state.

Bond order – half the number of electrons in bonding orbitals minus half the electrons in antibonding orbitals.

Bonding orbital – a molecular orbit lower in energy than any of the atomic orbitals from which it is derived; lends stability to a molecule or ion when populated with electrons.

Bonding pair – pair of electrons involved in a covalent bond.

Boron hydrides – binary compounds of boron and hydrogen.

Born-Haber cycle – a series of reactions (and the accompanying enthalpy changes) which, when summed, represent the hypothetical one-step reaction by which elements in their standard states are converted into crystals of ionic compounds (and the accompanying enthalpy changes).

Boyle's Law – at a constant temperature, the volume occupied by a definite mass of a gas is inversely proportional to the applied pressure.

Breeder reactor – a nuclear reactor that produces more fissionable nuclear fuel than it consumes.

Brønsted-Lowrey acid – a chemical species that donates a proton.

Brønsted-Lowrey base – a chemical species that accepts a proton.

Buffer solution – resists change in pH; contains either a weak acid and a soluble ionic salt of the acid or a weak base and a soluble ionic salt of the base.

Buret – a piece of volumetric glassware, usually graduated in 0.1 mL intervals, used to deliver solutions for titrations in a quantitative (drop-like) manner; also spelled *burette*.

C

Calorie – the amount of heat required to raise the temperature of one gram of water from 14.5 °C to 15.5 °C; 1 calorie = 4.184 joules.

Calorimeter – a device used to measure the heat transfer between a system and its surroundings.

Canal ray – a stream of positively charged particles (cations) that moves toward the negative electrode in cathode ray tubes; observed to pass through canals in the negative electrode.

Capillary – a tube having a very small inside diameter.

Capillary action – the drawing of a liquid up the inside of a small-bore tube when adhesive forces exceed cohesive forces; the depression of the surface of the liquid when cohesive forces exceed the adhesive forces.

Catalyst – a chemical compound used to change the rate (to speed or slow it) of a regenerated reaction (i.e., not consumed) at the end of the reaction.

Catenation – the bonding of atoms of the same element into chains or rings (i.e., the ability of an element to bond with itself).

Cathode – the electrode at which reduction occurs; in a cathode ray tube, the negative electrode.

Cathodic protection – protection of a metal (making a cathode) against corrosion by attaching it to a sacrificial anode of a more easily oxidized metal.

Cathode ray tube – a closed glass tube containing gas under low pressure, with electrodes near the ends and a luminescent screen near the positive electrode; produces cathode rays when a high voltage is applied.

Cation – a positive ion; an atom or group of atoms that has lost one or more electrons.

Cell potential – the potential difference, E_{cell}, between oxidation and reduction half-cells under nonstandard conditions; the force in a galvanic cell pulls electrons through a reducing agent to an oxidizing agent.

Central atom – an atom in a molecule or polyatomic ion bonded to more than one other atom.

Chain reaction – a reaction that, once initiated, sustains itself and expands; a reaction in which reactive species, such as radicals, are produced in more than one step, allowing these reactive species to propagate the chain reaction.

Charles' Law – at constant pressure, the volume occupied by a definite mass of gas is directly proportional to its absolute temperature.

Chemical bonds – the attractive forces holding atoms together in elements or compounds.

Chemical change – when one or more new substances are formed.

Chemical equation – description of a chemical reaction by placing the formulas of the reactants on the left of an arrow and the formulas of the products on the right.

Chemical equilibrium – a state of dynamic balance in which the rates of forward and reverse reactions are equal; there is no net change in concentrations of reactants or products while a system is at equilibrium.

Chemical kinetics – studies rates and mechanisms of chemical reactions and factors they depend on.

Chemical periodicity – the variations in properties of elements with their position in the periodic table.

Chemical reaction – the change of one or more substances into another or multiple substances.

Cloud chamber – a device for observing the paths of speeding particles as vapor molecules condense on them to form fog-like tracks.

Cobalt chloride paper – water test; water changes the color from blue to pink.

Coefficient of expansion – the ratio of the change in the length or the volume of a body to the original length or volume for a unit change in temperature.

Cohesive forces – the forces of attraction among particles of a liquid.

Colligative properties – physical properties of solutions that depend upon the number but not the kind of solute particles present.

Collision theory – the theory of reaction rates that states that effective collisions between reactant molecules must occur for the reaction to occur.

Colloid – a heterogeneous mixture in which solute-like particles do not settle (e.g., many kinds of milk).

Combination reaction – two substances (elements or compounds) combine to form one compound.

Combustible – classification of liquid substances that burn based on flashpoints; any liquid having a flashpoint at or above 37.8 °C (100 °F) but below 93.3 °C (200 °F), except any mixture having components with flashpoints of 93.3 °C (200 °F) or higher, the total of which makes up 99% or more of the volume of the mixture.

Combustion (or *burning*) – an exothermic reaction between an oxidant and fuel with heat and often light.

Common ion effect – suppression of ionization of a weak electrolyte by the presence in the same solution of a strong electrolyte containing one of the same ions as the weak electrolyte.

Complex ions – ions resulting from coordinating covalent bonds between simple ions and other ions or molecules.

Composition stoichiometry – describes the quantitative (mass) relationships among elements in compounds.

Compound – a substance of two or more chemically bonded elements in fixed proportions; can be decomposed into constituent elements.

Compressed gas – a single or mixture of gases having (in a container) an absolute pressure exceeding 40 psi at 21.1 °C (70 °F).

Compression – an area in a longitudinal wave where the particles are closer and pushed in.

Concentration – the amount of solute per unit volume, the mass of solvent or solution.

Condensation – the phase change from gas to liquid.

Condensed phases – the liquid and solid phases; phases in which particles interact strongly.

Condensed states – the solid and liquid states.

Conduction – heat transfer between substances in direct contact with each other (i.e., must be touching); when particles of a hotter substance vibrate, these molecules bump into nearby particles and transfer some energy.

Conduction band – a vacant or partially filled band of energy levels just higher in energy than a filled band; a band within which, or into which, electrons must be promoted to allow electrical conduction to occur in a solid.

Conductor – a material that allows electric flow more freely.

Conjugate acid-base pair – in Brønsted-Lowry terms, reactant and product that differ by a proton, H^+.

Conformations – structures of a compound that differ by the extent of rotation about a single bond.

Continuous spectrum – contains wavelengths in a specified region of the electromagnetic spectrum.

Control rods – rods of materials such as cadmium or boron steel that act as neutron absorbers (not merely moderators), used in nuclear reactors to control neutron fluxes and therefore fission rates.

Conjugated double bonds – double bonds separated from each other by one single bond $-C=C-C=C-$

Contact process – the industrial process for sulfur trioxide (SO_3) and sulfuric acid (H_2SO_4) production from sulfur dioxide (SO_2).

Convection – the physical flow of matter when heat flows by energized molecules from one place to another through the movement of fluids. Transfer of heat through a liquid or a gas occurs when molecules of the liquid or gas move and carry the heat.

Coordinate covalent bond – a covalent bond with shared electrons furnished by the same species; a bond between a Lewis acid and a Lewis base.

Coordination compound or complex – a compound containing coordinate covalent bonds.

Coordination number – the number of donor atoms coordinated to a metal; the number of nearest neighbors of an atom or ion in describing crystals.

Coordination sphere – the metal ion and its coordinating ligands, but no uncoordinated counter-ions.

Corrosion – oxidation of metals (e.g., rusting) in the presence of air and moisture.

Coulomb – the SI unit of electrical charge; unit symbol – C.

Covalent bond – a force of attraction (chemical bond) formed by the sharing of electron pairs between two atoms.

Covalent compounds – compounds made of two or more nonmetal atoms bonded by sharing valence electrons.

Critical mass – the minimum mass of a particular fissionable nuclide in a given volume required to sustain a nuclear chain reaction.

Critical point – the combination of critical temperature and critical pressure of a substance.

Critical pressure – the pressure required to liquefy a gas (vapor) at its *Critical temperature*.

Critical temperature – the temperature above which a gas cannot be liquefied; the temperature above which a substance cannot exhibit distinct gas and liquid phases.

Crystal – a solid packed with ions, molecules, or atoms in an orderly fashion.

Crystal field stabilization energy – a measure of the net energy of stabilization gained by a metal ion's nonbonding d electrons due to complex formation.

Crystal field theory – bonding in transition metal complexes in which ligands and metal ions are treated as point charges; a purely ionic model; ligand point charges represent the crystal (electrical) field perturbing the metal's d orbitals containing nonbonding electrons.

Crystal lattice – a pattern of arrangement of particles in a crystal.

Crystal lattice energy – the amount of energy that holds a crystal together; the energy change when a mole of solid forms from its constituent molecules or ions (for ionic compounds) in their gaseous state (consistently negative).

Crystalline solid – a solid characterized by a regular, ordered arrangement of particles.

Curie (Ci) – the basic unit to describe the intensity of radioactivity in a sample of material; one curie equals 37 billion disintegrations per second or approximately the amount of radioactivity given off by 1 gram of radium.

Current – a flow of charged particles, such as electrons or ions, moving through an electrical space or conductor. It is measured as the net rate of flow of electric charge; the unit is an Ampere (A).

Cuvette – glassware used in spectroscopic experiments; usually made of plastic, glass, or quartz, and should be as clean and transparent as possible.

Cyclotron – a device for accelerating charged particles along a spiral path.

D

Dalton's Law (or the *law of partial pressures*) – the pressure exerted by a mixture of gases is the sum of the partial pressures of the individual gases.

Daughter nuclide – nuclide produced in nuclear decay.

Debye – the unit used to express dipole moments.

Degenerate – in orbitals, describes orbitals of the same energy.

Deionization – the removal of ions; in the case of water, mineral ions such as sodium, iron, and calcium.

Deliquescence – substances that absorb water from the atmosphere to form liquid solutions.

Delocalization – in reference to electrons, bonding electrons distributed among more than two atoms bonded; occurs in species that exhibit resonance.

Density – mass per unit volume; $D = M \times V$.

Deposition – settling particles within a solution; the direct solidification of vapor by cooling; see *sublimation*.

Derivative – a compound that can be imagined arising from a parent compound by replacing one atom with another atom or group of atoms; used extensively in organic chemistry to identify compounds.

Detergent – a soap-like emulsifier with sulfate, SO_3, or a phosphate group instead of carboxylate group.

Deuterium – an isotope of hydrogen whose atoms are twice as massive as ordinary hydrogen; deuterium atoms contain a proton and a neutron in the nucleus.

Dextrorotatory – refers to an optically active substance that rotates plane-polarized light clockwise, also known as *dextro*.

Diagonal similarities – chemical similarities in the Periodic Table of Elements of Period 2 to elements of Period 3 one group to the right, especially evident toward the left of the periodic table.

Diamagnetism – weak repulsion by a magnetic field.

Differential Scanning Calorimetry (DSC) – a technique for measuring temperature, direction, and magnitude of thermal transitions in a sample material by heating/cooling and comparing the amount of energy required to maintain its rate of temperature increase or decrease with an inert reference material under similar conditions.

Differential Thermal Analysis (DTA) – a technique for observing the temperature, direction, and magnitude of thermally induced transitions in a material by heating/cooling a sample and comparing its temperature with an inert reference material under similar conditions.

Differential thermometer – a thermometer used to measure minimal temperature changes accurately.

Dilution – the process of reducing the concentration of a solute in a solution, usually by mixing with more solvent.

Dimer – a molecule formed by combining two smaller (identical) molecules.

Dipole – electric or magnetic separation of charge; charge separation between two covalently bonded atoms.

Dipole-dipole interactions – attractive electrostatic forces between polar molecules (i.e., between molecules with permanent dipoles).

Dipole moment – the product of the distance separating opposite charges of equal magnitude; a measure of the polarity of a bond or molecule; a measured dipole refers to the dipole moment of an entire molecule.

Dispersing medium – the solvent-like phase in a colloid.

Dispersed phase – the solute-like species in a colloid.

Displacement reactions – reactions in which one element displaces another from a compound.

Disproportionation reactions – redox reactions in which the oxidizing agent and the reducing agent are the same species.

Dissociation – in an aqueous solution, the process by which a solid ionic compound separates into its ions.

Dissociation constant – equilibrium constant for dissociating a complex ion into a simple ion and coordinating species (ligands).

Dissolution or solvation – the spread of ions in a monosaccharide.

Distilland – the material in a distillation apparatus that is to be distilled.

Distillate – the material in a distillation apparatus collected in the receiver.

Distillation – separating a liquid mixture into its components based on differences in boiling points; the process in which components of a mixture are separated by boiling away the more volatile liquid; the vaporization of a liquid by heating and then the condensation of the vapor by cooling.

Domain – a cluster of atoms in a ferromagnetic substance, which align in the same direction in the presence of an external magnetic field.

Donor atom – a ligand atom whose electrons are shared with a Lewis acid.

d-orbitals – beginning in the third energy level, a set of five degenerate orbitals per energy level, higher in energy than *s* and *p* orbitals of the same energy level.

Dosimeter – a small, calibrated electroscope worn by laboratory personnel to measure incident ionizing radiation or chemical exposure.

Double bond – covalent bond resulting from the sharing of four electrons (two pairs) between two atoms.

Double salt – a solid consisting of two co-crystallized salts.

Doublet – two peaks or bands of about equal intensity appearing close on a spectrogram.

Downs cell – an electrolytic cell for the commercial electrolysis of molten sodium chloride.

DP number – the degree of polymerization; the average number of monomer units per polymer unit.

Dry cells (voltaic cells) – ordinary batteries for appliances (e.g., flashlights, radios).

Dumas method – a method used to determine the molecular weights of volatile liquids.

Dynamic equilibrium – an equilibrium in which the processes occur continuously with no net change.

E

Earth metal – highly reactive elements in group IIA of the periodic table (includes beryllium, magnesium, calcium, strontium, barium, and radium); see *alkaline earth metal*.

Effective collisions – a collision between molecules resulting in a reaction, one in which the molecules collide with proper relative orientations and sufficient energy to react.

Effective molality – the sum of the molalities of solute particles in a solution.

Effective nuclear charge – the nuclear charge experienced by the outermost electrons of an atom; the actual nuclear charge minus the effects of shielding due to inner-shell electrons (e.g., a set of $d_{x^2-y^2}$ and d_{z^2} orbitals); those d orbitals within a set with lobes directed along the x, y, and z axes.

Electrical conductivity – the measure of how easily an electric current can flow through a substance.

Electric charge – a measured property (coulombs) that determines electromagnetic interaction.

Electrochemical cell – using a chemical reaction, an electromotive force is generated.

Electrochemistry – the study of chemical changes produced by electrical current and electricity production by chemical reactions.

Electrodes – surfaces upon which oxidation and reduction half-reactions occur in electrochemical cells; a conductor dips into an electrolyte and allows the electrons to flow to and from the electrolyte.

Electrode potentials – potentials, E, of half-reactions as reductions vs. standard hydrogen electrode.

Electrolysis – 1) occurs in electrolytic cells; chemical decomposition occurs by passing an electric current through a solution containing ions. 2) producing a chemical change using electricity; used to split up water into H and O_2.

Electrolyte – a substance (i.e., anions, cations) which, when dissolved in water, conducts electricity. An ionic solution that conducts a certain amount of current is categorized into two types: weak and strong.

Electrolytic cells – electrochemical cells in which electrical energy causes nonspontaneous redox reactions to occur (i.e., forced to occur by applying an outside source of electrical energy).

Electrolytic conduction – electrical current passes through ions in a solution or pure liquid.

Electromagnetic radiation – energy propagated using electric and magnetic fields that oscillate in directions perpendicular to the direction of travel of the energy; a type of wave that can go through vacuums as well as material; classified as a "self-propagating wave."

Electromagnetism – fields of an electric charge and electric properties that change how particles move and interact.

Electromotive force – a device that gains energy as electric charges pass through it.

Electromotive series – the relative order of tendencies for elements and their simple ions to act as oxidizing or reducing agents; also known as the "activity series."

Electron – a subatomic particle having a mass of 0.00054858 amu and a charge of -1.

Electron affinity – the energy absorbed in the process in which an electron is added to a neutral isolated gaseous atom to form a gaseous ion with a $1-$ charge; it has a negative value if energy is released.

Electron configuration – the specific distribution of electrons in atomic orbitals of atoms or ions.

Electron-deficient compounds – at least one atom (other than H) shares fewer than eight electrons.

Electron shells – an orbital around an atom's nucleus with a fixed number of electrons (usually 2 or 8).

Electronic transition – the transfer of an electron from one energy level to another.

Electronegativity – a measure of the relative tendency of an atom to attract electrons to itself when chemically combined with another atom.

Electronic geometry – the geometric arrangement of orbitals containing the shared and unshared electron pairs surrounding the central atom or polyatomic ion.

Electrophile – positively charged or electron-deficient.

Electrophoresis – separating ions by migration rate and direction of migration in an electric field.

Electroplating – a metal is covered with another metal layer using electricity; plating a metal onto a (cathodic) surface by electrolysis.

Element – a substance that cannot be decomposed into simpler substances by chemical means; defined by its *atomic number*. A substance that cannot be split into simpler substances by chemical means.

Eluant (or eluent) – the solvent used in the process of elution, as in liquid chromatography.

Eluate – a solvent (or mobile phase) which passes through a chromatographic column and removes the sample components from the stationary phase.

Emission spectrum – the emission of electromagnetic radiation by atoms (or other species) resulting from electronic transitions from higher to lower energy states.

Empirical formula – the simplest whole-number ratio of atoms of each element present in a compound; also known as the *simplest formula*.

Emulsifying agent – a substance that coats the particles of the dispersed phase and prevents coagulation of colloidal particles; an emulsifier.

Emulsion – colloidal suspension of a liquid in a liquid.

Endothermic – describes processes that absorb heat energy (H).

Endothermicity – the absorption of heat by a system as the process occurs.

Endpoint – the point at which an indicator changes color and titration stops.

Energy – a system's ability to do work.

Enthalpy (*H*) – the heat content of a specific amount of substance; $E = PV$.

Entropy *(S)* – a thermodynamic state or property that measures the degree of disorder (i.e., randomness) of a system; the amount of energy not available for work in a closed thermodynamic system (usually denoted by S).

Enzyme – a protein that acts as a catalyst in biological systems.

Equation of state – an expression that describes the behavior of matter in a given state; the van der Waals equation describes the behavior of the gaseous state.

Equilibrium or chemical equilibrium – a state of dynamic balance with the rates of forward and reverse reactions equal; the state of a system where neither forward nor reverse reaction is thermodynamically favored.

Equilibrium constant – a quantity characterizing equilibrium position for a reversible reaction; its magnitude is equal to mass action expression at equilibrium; equilibrium, "K," varies with temperature.

Equivalence point – the point when chemically equivalent amounts of reactants have reacted.

Equivalent weight – an oxidizing or reducing agent whose mass gains (oxidizing agents) or loses (reducing agents) 6.022×10^{23} electrons in a redox reaction.

Evaporation – vaporization of a liquid below its boiling point.

Evaporation rate – the rate at which a particular substance vaporizes (evaporate) compared to the rate of a known substance, such as ethyl ether, especially useful for health and fire-hazard considerations.

Excited state – any state other than the ground state of an atom or molecule; see *ground state*.

Exothermic – describes processes that release heat energy (*H*).

Exothermicity – the release of heat by a system as a process occurs.

Explosive – a chemical or compound that causes a sudden, almost instantaneous release of pressure, gas, heat, and light when subjected to sudden shock, pressure, high temperature, or applied potential.

Explosive limits – the range of concentrations over which a flammable vapor mixed with the proper ratios of air will ignite or explode if a source of ignition is provided.

Extensive property – a property that depends upon the amount of material in a sample.

Extrapolate – to estimate the value of a result outside the range of a series of known values; a technique used in the standard additions calibration procedure.

F

Faraday constant – a unit of electrical charge widely used in electrochemistry and equal to ~ 96,500 coulombs; represents 1 mole of electrons, or the Avogadro number of electrons: 6.022×10^{23} electrons.

Faraday's law of electrolysis – a two-part law that Michael Faraday published about electrolysis. 1. the mass of a substance altered at an electrode during electrolysis is directly proportional to the quantity of electricity transferred at that electrode. 2. the mass of an elemental material altered at an electrode is directly proportional to the element's equivalent weight; one equivalent weight of a substance is produced at each electrode during the passage of 96,487 coulombs of charge through an electrolytic cell.

Fast neutron – a neutron ejected at high kinetic energy in a nuclear reaction.

Ferromagnetism – the ability of a substance to become permanently magnetized by exposure to an external magnetic field.

Flashpoint – the temperature at which a liquid yields enough flammable vapor to ignite; various recognized industrial testing methods exist; therefore, the method used must be specified.

Fluorescence – absorption of high-energy radiation and subsequent emission of visible light.

First Law of Thermodynamics – the amount of energy in the universe is constant (i.e., energy is neither created nor destroyed in ordinary chemical reactions and physical changes); known as the Law of Conservation of Energy.

Fluids – substances that flow freely; gases and liquids.

Flux – a substance added to react with the charge or a product of its reduction; in metallurgy, it is usually added to lower the melting point.

Foam – colloidal suspension of a gas in a liquid.

Formal charge – a method of counting electrons in a covalently bonded molecule or ion; it counts bonding electrons as though they were equally shared between the two atoms.

Formula – a combination of symbols that indicates the chemical composition of a substance.

Formula unit – the smallest repeating unit of a substance; the molecule for nonionic substances.

Formula weight – the mass of one formula unit of a substance in atomic mass units.

Fossil fuels – formed from the remains of plants and animals that lived millions of years ago.

Fractional distillation – when a fractioning column is used in a distillation apparatus to separate the components of a liquid mixture with different boiling points.

Fractional precipitation – removal of some ions from a solution by precipitation while leaving other ions with similar properties in the solution.

Free energy change – the indicator of the spontaneity of a process at constant T and P (e.g., if ΔG is negative, the process is spontaneous).

Free radical – a highly reactive chemical species carrying no charge and having a single unpaired electron in an orbital.

Freezing – phase transition from liquid to solid.

Freezing point depression – the decrease in the freezing point of a solvent caused by the presence of a solute.

Frequency – the number of repeating points on a wave that passes a given observation point per unit time; the unit is 1 hertz = 1 cycle per 1 second.

Fuel – any substance that burns in oxygen to produce heat.

Fuel cells – a voltaic cell that converts the chemical energy of a fuel and an oxidizing agent directly into electrical energy continuously.

G

Gamma (γ) ray – a highly penetrating type of nuclear radiation similar to X-ray radiation, except that it comes from within the nucleus of an atom and has higher energy; energy-wise, very similar to cosmic rays, except that cosmic rays originate from outer space.

Galvanic cell – a battery made from an electrochemical cell with two metals connected by a salt bridge.

Galvanizing – placing a thin layer of zinc on a ferrous material to protect the underlying surface from corrosion.

Gangue – sand, rock, and other impurities surrounding the mineral of interest in an ore.

Gas – a state of matter in which the particles have no definite shape or volume, though they fill their container.

Gay-Lussac's Law – the expression used for each of the two relationships named after the French chemist Joseph Louis Gay-Lussac concerning the properties of gases; more usually applied to his law of combining volumes.

Geiger counter – a gas-filled tube that discharges electrically when ionizing radiation passes through it.

Gel – colloidal suspension of a solid dispersed in a liquid; a semi-rigid solid.

Gibbs (free) energy – the thermodynamic state function of a system that indicates the amount of energy available for the system to do useful work at constant T and P; a value that indicates the spontaneity of a reaction (usually denoted by G).

Graham's Law – the rates of effusion of gases are inversely proportional to the square roots of their molecular weights or densities.

Ground state – the lowest energy state or most stable state of an atom, molecule, or ion; see *excited state*.

Group – a vertical column in the periodic table; known as a family.

H

Haber process – a process for the catalyzed industrial production of ammonia from N_2 and H_2 at high temperature and pressure.

Half-cell – the compartment in which the oxidation or reduction half-reaction occurs in a voltaic cell.

Half-life – the time required for half of a reactant to be converted into product(s); the time required for half of a given sample to undergo radioactive decay.

Half-reaction – the oxidation or the reduction part of a redox reaction.

Halogens – group VIIA elements: F, Cl, Br, I; halogens are nonmetals.

Hard water – water high in dissolved minerals that makes it difficult to form lather with soap.

Heat – a form of energy that flows between two samples of matter because of temperature differences.

Heat capacity – the amount of heat required to raise the temperature of a mass one degree Celsius.

Heat of condensation – the amount of heat that must be removed from one gram of vapor at its condensation point to condense the vapor with no change in temperature.

Heat of crystallization – the amount of heat that must be removed from one gram of a liquid at its freezing point to freeze it with no change in temperature.

Heat of fusion – the amount of heat required to melt one gram of a solid at its melting point with no change in temperature; usually expressed in J/g; the molar heat of fusion is the amount of heat required to melt one mole of a solid at its melting point with no change in temperature and is usually expressed in kJ/mol.

Heat of solution – the amount of heat absorbed in forming a solution that contains one mole of the solute; the value is positive if heat is absorbed (endothermic) and negative if heat is released (exothermic).

Heat of vaporization – the amount of heat required to vaporize one gram of a liquid at its boiling point with no change in temperature; usually expressed in J/g; the molar heat of vaporization is the amount of heat required to vaporize one mole of liquid at its boiling point with no change in temperature and is usually expressed as ion kJ/mol.

Heisenberg uncertainty principle – states that it is impossible to determine the momentum and position of an electron simultaneously with absolute accuracy.

Henry's Law – the gas pressure above a solution is proportional to concentration of the gas in solution.

Hess' Law of heat summation – the enthalpy change for a reaction is the same whether it occurs in one step or a series of steps.

Heterogeneous catalyst – exists in a different phase (solid, liquid, or gas) from the reactants; a contact catalyst.

Heterogeneous equilibria – equilibria involving species in more than one phase.

Heterogeneous mixture – a mixture that does not have uniform composition and properties throughout.

Heteronuclear – consisting of different elements.

High spin complex – crystal field designation for an outer orbital complex; t_{2g} and e_g orbitals are singly occupied before pairing occurs.

Homogeneous catalyst – in the same phase (solid, liquid, or gas) as the reactants.

Homogeneous equilibria – when *reagents* and *products* are in the same phase (gases, liquids, or solids).

Homogeneous mixture – a mixture which has uniform composition and properties throughout.

Homologous series – compounds with each member differing from the next by a specific number and kind of atoms.

Homonuclear – consisting of only one element.

Hund's rule – single electrons must occupy orbitals of a given sublevel before pairing begins; see *Aufbau* (or *building up*) *principle*.

Hybridization – mixing atomic orbitals to form a new set of atomic orbitals with the same electron capacity and properties and energies intermediate between the original unhybridized orbitals.

Hydrate – a solid compound that contains a definite percentage of bound water.

Hydrate isomers – crystalline complexes that differ in whether water exists inside or outside the coordination sphere.

Hydration – the reaction of a substance with water.

Hydration energy – the energy change accompanying the hydration of a mole of gas and ions.

Hydride – a binary compound of hydrogen.

Hydrocarbons – compounds that contain only carbon and hydrogen.

Hydrogen bond – a relatively strong dipole-dipole interaction (but still considerably weaker than the covalent or ionic bonds) between molecules containing hydrogen directly bonded to a small, highly electronegative atom, such as N, O or F.

Hydrogenation – the reaction in which hydrogen adds across a double or triple bond.

Hydrogen-oxygen fuel cell – hydrogen is the fuel (reducing agent), and oxygen is the oxidizing agent.

Hydrolysis – the reaction of a substance with water or its ions.

Hydrolysis constant – an equilibrium constant for a hydrolysis reaction.

Hydrometer – a device used to measure the densities of liquids and solutions.

Hydrophilic colloids – colloidal particles that repel water molecules.

I

Ideal gas – a hypothetical gas that obeys the postulates of the Kinetic Molecular Theory.

Ideal gas law – the product of pressure and the volume of an ideal gas is directly proportional to the number of moles of the gas and the absolute temperature. $PV = nRT$

Ideal solution – obeys Raoult's Law strictly.

Immiscible liquids – do not mix to form a solution (e.g., oil and water).

Indicators – for acid-base titrations, organic compounds that exhibit different colors in solutions of different acidities, determine the point at which the reaction between two solutes is complete.

Inert pair effect – characteristic of the post-transition minerals; the tendency of the electrons in the outermost atomic *s* orbital to remain un-ionized or unshared in compounds of post-transition metals.

Inhibitory catalyst – an inhibitor; a catalyst that decreases the rate of reaction.

Inner orbital complex – valence bond designation for a complex in which the metal ion utilizes d orbitals for one shell inside the outermost occupied shell in its hybridization.

Inorganic chemistry – a part of chemistry concerned with inorganic (non-carbon-based) compounds.

Insulator – a material that resists the flow of electric current or heat transfer; it does not allow heat to flow easily.

Insoluble compound – a substance that will not dissolve in a solvent, even after mixing.

Integrated rate equation – an expression giving the concentration of a reactant remaining after a specified time; it has a different mathematical form for different orders of reactants.

Intermolecular forces – forces between individual particles (atoms, molecules, ions) of a substance.

Ion – a molecule that has gained or lost electrons; an atom or a group of atoms carries an electric charge (Na^+).

Ion product for water – equilibrium constant for water ionization; $Kw = [H_3O^+]\cdot[^-OH] = 1.00 \times 10^{-14}$ at 25 °C.

Ionic bond – the electrostatic attraction between oppositely charged ions, resulting from a transfer of electrons.

Ionic bonding – chemical bonding resulting from transferring electrons from one atom or group.

Ionic compounds – compounds containing predominantly ionic bonding.

Ionic geometry – arrangement of atoms (not lone pairs of electrons) about the central atom of a polyatomic ion.

Ionization – the breaking up of a compound into separate ions; in an aqueous solution, the process by which a molecular compound reacts with water and forms ions.

Ionization constant – equilibrium constant for the ionization of a weak electrolyte.

Ionization energy – the minimum amount of energy required to remove the most loosely held electron of an isolated gaseous atom or ion.

Ion exchange – a method of removing hardness from water, it replaces the positive ions that cause the hardness with H^+ ions.

Ionization isomers – result from the interchange of ions inside and outside the coordination sphere.

Isoelectric – having the same electronic configurations.

Isomers – different substances with the same formula.

Isomorphous – refers to crystals having the same atomic arrangement.

Isotopes – two or more forms of atoms of the same element with different masses; atoms containing the same number of protons but different numbers of neutrons.

IUPAC – acronym for "International Union of Pure and Applied Chemistry."

J

Joule (J) – a unit of energy in the SI system; one joule is $1 \; kg\cdot m^2/s^2$, which is 0.2390 calories.

K

K capture – absorption of a K shell (n = 1) electron by a proton as it is converted to a neutron.

Kelvin – a unit of measure for temperature based upon an absolute scale.

Kinetics – a subfield of chemistry specializing in reaction rates.

Kinetic energy (*KE*) – energy that matter processes through its motion.

Kinetic Molecular Theory – a theory that attempts to explain macroscopic observations on gases in microscopic or molecular terms.

L

Lanthanides – elements 57 (lanthanum) through 71 (lutetium); grouped because of their similar behavior in chemical reactions.

Lanthanide contraction – a decrease in the radii of the elements following the lanthanides compared to what would be expected if there were no *f*-transition metals.

Latent heat – the energy absorbed or released when a substance changes state without changing temperature.

Lattice – a unique arrangement of atoms or molecules in a crystalline liquid or solid.

Law of combining volumes (Gay-Lussac's Law) – at constant temperature and pressure, the volumes of reacting gases (and any gaseous products) can be expressed as ratios of small whole numbers.

Law of conservation of energy – energy cannot be created or destroyed; it can only change form.

Law of conservation of matter – there is no detectable change in the quantity of matter during an ordinary chemical reaction.

Law of conservation of matter and energy – the amount of matter and energy in the universe is fixed.

Law of definite proportions (law of constant composition) – different samples of a pure compound contain the same elements in the same proportions by mass.

Law of partial pressures (or *Dalton's Law*) – the pressure exerted by a mixture of gases is the sum of the partial pressures of the individual gases.

Laws of thermodynamics – physical laws which define quantities of thermodynamic systems describe how they behave and (by extension) set certain limitations, such as perpetual motion.

Lead storage battery – a secondary voltaic cell used in most automobiles.

Leclanche cell – a common type of *dry cell*.

Le Chatelier's principle – states that a system at equilibrium, or striving to attain equilibrium, responds in such a way as to counteract any stress placed upon it; if stress (change of conditions) is applied to a system at equilibrium, the system will shift in the direction that reduces stress.

Leveling effect – acids stronger than the acid characteristic of the solvent react with the solvent to produce that acid; a similar statement applies to bases. The strongest acid (base) that can exist in a given solvent is the acid (base) characteristic of the solvent.

Levorotatory (or *levo*) – an optically active substance rotates plane-polarized light counterclockwise.

Lewis acid – any species that can accept a share in an electron pair.

Lewis base – any species that can make available a pair of electrons.

Lewis dot formula (electron dot formula) – representation of a molecule, ion, or formula unit by showing atomic symbols and only outer shell electrons.

Ligand – a Lewis base in a coordination compound.

Light – that portion of the electromagnetic spectrum visible to the naked eye; known as "visible light."

Limiting reactant – a substance that stoichiometrically limits the number of products formed.

Linear accelerator – a device used for accelerating charged particles along a straight line path.

Line spectrum – an atomic emission or absorption spectrum.

Linkage isomers – a particular ligand bonds to a metal ion through different donor atoms.

Liquid – a state of matter which takes the shape of its container.

Liquid aerosol – colloidal suspension of a liquid in a gas.

London dispersion forces – very weak and very short-range attractive forces between short-lived temporary (induced) dipoles; known as "dispersion forces."

Lone pair – a pair of electrons residing on one atom and not shared by other atoms; an unshared pair.

Low spin complex – crystal field designation for an inner orbital complex; contains electrons paired t_{2g} orbitals before e_g orbitals are occupied in octahedral complexes.

Lubricant – a substance capable of reducing friction (i.e., a force that opposes the direction of motion).

M

Magnetic field – a space around a magnet where magnetism can be detected.

Magnetic quantum number – quantum mechanical solution to a wave equation designating the orbital within a given set (s, p, d, f) in which an electron resides.

Manometer – a two-armed barometer.

Mass (m) – a measure of the amount of matter in an object; mass is usually measured in grams or kilograms.

Mass action expression – for a reversible reaction, aA + bB cC + dD; the product of the concentrations of the products (species on the right), each raised to the power corresponding to its coefficient in the balanced chemical equation, divided by the product of the concentrations of reactants (species on the left), each raised to the power corresponding to its coefficient in the balanced equation; at equilibrium the mass action expression is equal to K.

Mass deficiency – the amount of matter converted into energy when an atom forms from constituent particles.

Mass number – the sum of the numbers of protons and neutrons in an atom; an integer.

Mass spectrometer – an instrument that measures the charge-to-mass ratio of charged particles.

Matter – anything that has mass and occupies space.

Mechanism – the sequence of steps by which reactants are converted into products.

Melting point – the temperature at which liquid and solid coexist in equilibrium.

Meniscus – the shape assumed by the surface of a liquid in a cylindrical container.

Melting – the phase change from a solid to a liquid.

Metal – a chemical element that is a good conductor of electricity and heat and forms cations and ionic bonds with nonmetals; elements below and to the left of the stepwise division (metalloids) in the upper right corner of the periodic table; about 80% of known elements are metals.

Metallic bonding – bonding within metals due to the electrical attraction of positively charged metal ions for mobile electrons that belong to the crystal.

Metallic conduction – conduction of electrical current through a metal or along a metallic surface.

Metalloid – a substance with the properties of metals and nonmetals (B, Al, Si, Ge, As, Sb, Te, Po and At).

Metathesis reactions – reactions in which two compounds react to form two new compounds, with no changes in oxidation number; reactions in which the ions of two compounds exchange partners.

Method of initial rates – method of determining the rate-law expression by carrying out a reaction with different initial concentrations and analyzing the resultant changes in initial rates.

Methylene blue – a heterocyclic aromatic chemical compound with the molecular formula $C_{16}H_{18}N_3SCl$.

Miscible liquids – mix to form a solution (e.g., alcohol and water).

Miscibility – the ability of one liquid to mix with (dissolve in) another liquid.

Mixture – two or more different substances mingled together but not chemically combined. A sample of matter composed of two or more substances, each of which retains its identity and properties.

Moderator – a substance (e.g., deuterium, oxygen, paraffin) that slows fast neutrons upon collision.

Molality – a concentration expressed as the number of moles of solute per kilogram of solvent.

Molarity – the number of moles of solute per liter of solution.

Molar solubility – the number of moles of a solute that dissolves to produce a liter of a saturated solution.

Mole – a measurement of an amount of substance; a single mole contains approximately 6.022×10^{23} units or entities; abbreviated mol.

Molecule – a chemically bonded number of electrically neutral atoms.

Molecular equation – a chemical reaction in which formulas are written as if substances existed as molecules; only complete formulas are used.

Molecular formula – indicates the number of atoms present in a molecule of a molecular substance.

Molecular geometry – the arrangement of atoms (not lone pairs of electrons) around a central atom of a molecule or polyatomic ion.

Molecular orbital (*MO*) – resulting from the overlap and mixing of atomic orbitals on different atoms (i.e., a region where an electron can be found in a molecule, as opposed to an atom); an MO belongs to the molecule.

Molecular orbital theory – a theory of chemical bonding based upon postulated molecular orbitals.

Molecular weight – the mass of one molecule of a nonionic substance in atomic mass units.

Molecule – the smallest particle of a compound capable of stable, independent existence.

Mole fraction – the number of moles of a component in a mixture divided by the number of moles in the mixture.

Monoprotic acid – can form only one hydronium ion per molecule; may be strong or weak.

Mother nuclide – nuclide that undergoes nuclear decay.

N

Native state – refers to the occurrence of an element in an uncombined or free state in nature.

Natural radioactivity – spontaneous decomposition of an atom.

Neat – conditions with a liquid reagent or gas performed with no added solvent or co-solvent.

Nernst equation – corrects standard electrode potentials for nonstandard conditions.

Net ionic equation – results from canceling spectator ions and eliminating brackets from a total ionic equation.

Neutralization – the reaction of an acid with a base to form a salt and water; usually, the reaction of hydrogen ions with hydroxide ions to form water molecules.

Neutrino – a particle that travels at speeds close to the speed of light; created due to radioactive decay.

Neutron – a neutral unit or subatomic particle with no net charge and a mass of 1.0087 amu.

Nickel-cadmium cell (NiCd battery) – a dry cell where the anode is Cd, the cathode is NiO_2, and the electrolyte is basic.

Nitrogen cycle – the complex series of reactions by which nitrogen is slowly but continually recycled in the atmosphere, lithosphere, and hydrosphere.

Noble gases – elements of the periodic Group 0; He, Ne, Ar, Kr, Xe, Rn; known as "rare gases;" formerly called "inert gases."

Nodal plane – a region in which the probability of finding an electron is zero.

Nonbonding orbital – a molecular orbital derived only from an atomic orbital of one atom; it lends neither stability nor instability to a molecule or ion when populated with electrons.

Nonelectrolyte – a substance whose aqueous solutions do not conduct electricity.

Nonmetal – an element that is not metallic.

Nonpolar bond – a covalent bond in which electron density is symmetrically distributed.

Nuclear – of or about the atomic nucleus.

Nuclear binding energy – the energy equivalent of the mass deficiency; the energy released in forming an atom from the subatomic particles.

Nuclear fission – when a heavy nucleus splits into nuclei of intermediate masses, and protons are emitted.

Nuclear magnetic resonance spectroscopy – a technique that exploits the magnetic properties of specific nuclei; helps identify unknown compounds.

Nuclear reaction – involves a change in the composition of a nucleus and can emit or absorb a tremendous amount of energy.

Nuclear reactor – a system in which controlled nuclear fission reactions generate heat energy on a large scale, subsequently converted into electrical energy.

Nucleons – particles comprising the nucleus: protons and neutrons.

Nucleus – the tiny and dense, positively charged center of an atom containing protons and neutrons, as well as other subatomic particles; the net charge is positive.

Nuclides – refers to different atomic forms of elements; in contrast to isotopes, which refer only to different atomic forms of a single element.

Nuclide symbol – designation for an atom A/Z E, in which E is the symbol of an element, Z is its atomic number, and A is its mass number.

Number density – a measure of the concentration of countable objects (e.g., atoms, molecules) in space; the number per volume.

O

Octahedral – molecules and polyatomic ions with one atom in the center and six atoms at the corners of an octahedron.

Octane number – a number that indicates how smoothly a gasoline burns.

Octet rule – during bonding, atoms tend to reach an electron arrangement with eight electrons in the outermost shell. Many representative elements attain at least a share of eight electrons in their valence shells when they form molecular or ionic compounds; there are some limitations.

Open sextet – species with only six electrons in the highest energy level of the central element (many Lewis acids).

Orbital – may refer to an atomic orbital or a molecular orbital.

Organic chemistry – the chemistry of substances that contain carbon-hydrogen bonds.

Organic compound – substances that contain carbon.

Osmosis – when solvent molecules pass through a semi-permeable membrane from a dilute solution into a more concentrated solution.

Osmotic pressure – the hydrostatic pressure produced on the surface of a semi-permeable membrane.

Outer orbital complex – valence bond designation for a complex in which the metal ion utilizes d orbitals in the outermost (occupied) shell in hybridization.

Overlap – the interaction of orbitals on different atoms in the same region of space.

Oxidation – the addition of oxygen or the loss of electrons. An algebraic increase in the oxidation number may correspond to a loss of electrons.

Oxidation numbers – quantitative values used as mechanical aids in writing formulas and balancing equations; for single-atom ions, they correspond to the charge on the ion; more electronegative atoms are assigned negative oxidation numbers, known as *oxidation states*.

Oxidation-reduction reactions – reactions in which oxidation and reduction occur; known as *redox reactions*.

Oxide – a binary compound of oxygen.

Oxidizing agent – the substance that oxidizes another substance and is reduced.

P

Pairing – a favorable interaction of two electrons with opposite m values in the same orbital.

Pairing energy – the energy required to pair two electrons in the same orbital.

Paramagnetism – attraction toward a magnetic field, stronger than diamagnetism but still weak compared to ferromagnetism.

Partial pressure – the force exerted by one gas in a mixture of gases.

Particulate matter – fine, divided solid particles suspended in polluted air.

Pauli exclusion principle – no electrons in the same atom may have identical sets of four quantum numbers.

Percentage ionization – the percentage of the weak electrolyte that will ionize in a solution of given concentration.

Percent by mass – 100% times the actual yield divided by the theoretical yield.

Percent composition – the mass percent of each element in a compound.

Percent purity – the percent of a specified compound or element in an impure sample.

Period – the elements in a horizontal row of the periodic table.

Periodicity – regular periodic variations of properties of elements with their atomic number (and position in the periodic table).

Periodic Law – the properties of the elements are periodic functions of their atomic numbers.

Periodic table – an arrangement of elements by increasing atomic numbers, emphasizing periodicity.

Peroxide – a compound with oxygen in –1 oxidation state; metal peroxides contain the peroxide ion, O_2^{2-}.

pH – the measure of acidity (or basicity) of a solution; negative logarithm of the concentration (mol/L) of the H_3O^+ [H^+] ion; scale is commonly used over a range 0 to 14.

Phase diagram – shows the equilibrium temperature-pressure relationships for different phases of a substance.

pH scale – a range from 0 to 14. If the pH of a solution is 7, it is neutral; if the pH of a solution is less than 7, it is acidic; if the pH of a solution is greater than 7, it is basic.

Permanent hardness – hardness (relative to lathering soap) in water that cannot be removed by boiling; caused by calcium sulfate.

Photoelectric effect – emission of an electron from the surface of a metal caused by impinging electromagnetic radiation of specific minimum energy; the current increases with increasing radiation intensity.

Photon – a carrier of electromagnetic radiation of all wavelengths, such as gamma rays and radio waves; known as a *quantum of light*.

Physical change – when a substance changes from one physical state to another, but no substances with different compositions are formed; physical change may involve a phase change (e.g., melting, freezing) or another physical change, such as crushing a crystal or separating one volume of liquid into different containers; it does not produce a new substance.

Plasma – a physical state of matter that exists at extremely high temperatures in which molecules are dissociated, and most atoms are ionized.

Polar bond – a covalent bond with an unsymmetrical distribution of electron density.

Polarimeter – a device used to measure optical activity.

Polarization – the buildup of a product of oxidation or reduction of an electrode, preventing further reaction.

Polydentate – refers to ligands with more than one donor atom.

Polyene – a compound that contains more than one double bond per molecule.

Polymerization – the combination of many small molecules to form large molecules.

Polymer – a large molecule consisting of chains or rings of linked monomer units, usually characterized by high melting and boiling points.

Polymorphous – refers to substances that can crystallize in more than one crystalline arrangement.

Polyprotic acid – forms two or more H_3O_+ ions per molecule; often, at least one ionization step is weak.

Positron – a nuclear particle with the mass of an electron but opposite charge (positive).

Potential difference (or *voltage*) – the force that moves the electrons around circuit; unit is Volt (V).

Potential energy (*PE*) – energy stored in a body or a system due to its position in a force field or configuration.

Power – the rate at which energy is converted from one form to another; the unit is Watts (W). Power = voltage $\times$ current (P = VI).

Precipitate – an insoluble solid formed by mixing in solution the constituent ions of a slightly soluble solution.

Precision – how close the results of multiple experimental trials are; see *accuracy*.

Pressure – force per unit area; unit is Pascal (Pa).

Primary standard – a known high degree of purity substance that undergoes one invariable reaction with the other reactant of interest.

Primary voltaic cells – voltaic cells that cannot be recharged; no further chemical reaction is possible once the reactants are consumed.

Products – chemicals produced (from reactants) in a chemical reaction.

Proton – a subatomic particle having a mass of 1.0073 amu and a charge of +1, found atom's nucleus.

Protonation – the addition of a proton (H^+) to an atom, molecule, or ion.

Pseudobinaryionic compounds – contain more than two elements but are named like binary compounds.

Q

Quanta – the minimum amount of energy emitted by radiation.

Quantum mechanics – the study of how atoms, molecules, subatomic particles, etc., behave and are structured; a mathematical method of treating particles based on quantum theory, which assumes that energy (of small particles) is not infinitely divisible.

Quantum numbers – numbers that describe the energies of electrons in atoms; derived from quantum mechanical treatment.

Quarks – elementary particles and a fundamental constituent of matter, combining to form hadrons (i.e., protons and neutrons).

R

Radiation – 1) heat transfer through invisible rays, which travel outwards from the hot object without a medium. 2) high-energy particles or rays emitted during the nuclear decay processes.

Radical – an atom or group of atoms that contains one or more unpaired electrons; usually a very reactive species.

Radioactive dating – a method of dating ancient objects by determining the ratio of mother and daughter nuclides present in an object and relating the ratio to the object's age via half-life calculations.

Radioactive tracer – a small amount of radioisotope replacing a nonradioactive isotope of the element in a compound whose path (e.g., in the body) or whose decomposition products are monitored by detection of radioactivity; known as a "radioactive label."

Radioactivity – the spontaneous disintegration of atomic nuclei.

Raoult's Law – the vapor pressure of a solvent in an ideal solution decreases as its mole fraction decreases.

Rate-determining step – the slowest step in a mechanism; determines the overall reaction rate.

Rate-law expression – an equation relating the reaction rate to the concentrations of the reactants and the specific rate of the reaction.

Rate of reaction – the change in the concentration of a reactant or product per unit time.

Reactants – substances consumed in a chemical reaction; react together in a chemical reaction.

Reaction quotient – the mass action expression under any set of conditions (not necessarily equilibrium); its magnitude relative to K determines the direction in which the reaction must occur to establish equilibrium.

Reaction ratio – the relative amounts of reactants and products involved in a reaction; may be the ratio of moles, millimoles, or masses.

Reaction stoichiometry – describes the quantitative relationships among substances participating in chemical reactions.

Reactivity series (or activity series) – an empirical, calculated, and structurally analytical progression of a series of metals, arranged by "reactivity" from highest to lowest; used to summarize information about the reactions of metals with acids and water, double displacement reactions, and the extraction of metals from ores.

Reagent – a substance (or compound) added to a system to cause a chemical reaction or to visualize if a reaction occurs; the terms reactant and reagent are often used interchangeably; however, a *reactant* is more specifically a substance consumed during a chemical reaction.

Reducing agent – a substance that reduces another substance and is itself oxidized.

Reduction – the removal of oxygen or the gaining of electrons.

Resonance – the concept in which two or more equivalent dot formulas for the same arrangement of atoms (resonance structures) are necessary to describe the bonding in a molecule or ion.

Reverse osmosis – forcing solvent molecules to flow through a semi-permeable membrane from a concentrated solution into a dilute solution by applying greater hydrostatic pressure on the concentrated side than the osmotic pressure opposing it.

Reversible reaction – processes that do not go to completion and occur in the forward and reverse direction.

S

Saline solution – a general term for NaCl (i.e., sodium chloride) in water.

Salt – when a metal replaces the hydrogen of an acid.

Salts – ionic compounds composed of anions and cations.

Salt bridge – a U-shaped tube containing an electrolyte, connects the two half-cells of a voltaic cell.

Saturated solution –no more solute will dissolve at that temperature.

s-block elements – group 1 and 2 elements (alkali and alkaline metals), including hydrogen and helium.

Schrödinger equation – quantum state equation representing the behavior of an electron around an atom; describes the wave function of a physical system evolving.

Second Law of Thermodynamics – the universe tends toward a state of greater disorder in spontaneous processes.

Secondary standard – a solution that has been titrated against a primary standard; a standard solution.

Secondary voltaic cells – voltaic cells that can be recharged; original reactants can be regenerated by reversing the direction of the current flow.

Semiconductor – a substance that does not conduct electricity at low temperatures but will do so at higher temperatures.

Semi-permeable membrane – a thin partition between two solutions through which specific molecules can pass but others cannot.

Shielding effect – electrons in filled sets of s, p orbitals between the nucleus and outer shell electrons shield the outer shell electrons somewhat from the effect of protons in the nucleus; known as the "screening effect."

Sigma (σ) bonds – bonds resulting from the head-on overlap of atomic orbitals. The region of electron sharing is along and (cylindrically) symmetrical to the imaginary line connecting the bonded atoms.

Sigma orbital – molecular orbital resulting from the head-on overlap of two atomic orbitals.

Single bond – covalent bond resulting from the sharing of two electrons (one pair) between two atoms.

Sol – a suspension of solid particles in a liquid; artificial examples include sol-gels.

Solid – one of the states of matter, where the molecules are packed closely, resistance to movement/deformation, and volume change.

Solubility product constant – equilibrium constant for the dissolution of a slightly soluble compound.

Solubility product principle – the solubility product constant expression for a slightly soluble compound is the product of the concentrations of the constituent ions; each raised to the power that corresponds to the number of ions in one formula unit.

Solute – the dispersed (i.e., dissolved) phase of a solution; the solution is mixed into the solvent (e.g., NaCl in saline water).

Solution – a homogeneous mixture of multiple substances; comprised of solutes and solvents; a mixture of a solute (usually a solid) and a solvent (usually a liquid).

Solvation – the process by which solvent molecules surround and interact with solute ions or molecules.

Solvent – the dispersing medium of a solution (e.g., H_2O in saline water).

Solvolysis – the reaction of a substance with the solvent in which it is dissolved.

s-orbital – a spherically symmetrical atomic orbital; one per energy level.

Specific gravity – the ratio of the density of a substance to the density of water.

Specific heat – the amount of heat required to raise the temperature of one gram of substance 1 °C.

Specific rate constant – an experimentally determined (proportionality) constant; different for different reactions, and which changes only with temperature; k in the rate-law expression: Rate = k [A] $\times$ [B].

Spectator ions – ions in a solution that do not participate in a chemical reaction.

Spectral line – any of several lines corresponding to definite wavelengths of an atomic emission or absorption spectrum; marks the energy difference between two energy levels.

Spectrochemical series – arrangement of ligands in order of increasing ligand field strength.

Spectroscopy – the study of radiation and matter, such as X-ray absorption and emission spectroscopy.

Spectrum – display of component wavelengths (colors) of electromagnetic radiation.

Speed of light – the speed at which radiation travels through a vacuum (299,792,458 m/sec).

Square planar – describes molecules and polyatomic ions with one atom in the center and four atoms at the corners of a square.

Square planar complex – relationship with metal in the center of a square plane, with ligand donor atoms at each of the four corners.

Standard conditions for temperature and pressure (STP) – used to compare experimental results (25 °C and 100.000 kPa).

Standard electrodes – half-cells in which the oxidized and reduced forms of a species are present at the unit activity (1.0 M solutions of dissolved ions, 1.0 atm partial pressure of gases, pure solids, and liquids).

Standard electrode potential – by convention, the potential (Eo) of a half-reaction as a reduction relative to the standard hydrogen electrode when species are present at unit activity.

Standard entropy – the absolute entropy of a substance in its standard state at 298 K.

Standard molar enthalpy of formation – the amount of heat absorbed in forming one mole of a substance in a specified state from its elements in their standard states.

Standard molar volume – space occupied by 1 mole of ideal gas under standard conditions; 22.4 liters.

Standard reaction – a process where the numbers of moles of reactants in the balanced equation, in their standard states, are entirely converted to the numbers of moles of products in the balanced equation, at their standard state.

State of matter – a homogeneous, macroscopic phase (e.g., gas, plasma, liquid, solid) in increasing concentration.

Stoichiometry – quantitative relationships of elements and compounds undergoing chemical changes.

Strong electrolyte – a substance that conducts electricity well in a dilute aqueous solution.

Strong field ligand – a ligand that exerts a strong crystal or ligand electrical field and generally forms low-spin complexes with metal ions when possible.

Structural isomers – compounds that contain the same number and kinds of atoms, but with different geometries.

Subatomic particles – comprise an atom (e.g., protons, neutrons, electrons).

Sublimation – the direct vaporization of a solid by heating without passing through the liquid state; a phase transition from solid to gas.

Substance – any matter, specimens with the same chemical composition and physical properties.

Substitution reaction – a reaction in which another atom or group of atoms replaces an atom or a group of atoms.

Supercooled liquids – liquids that, when cooled, apparently solidify but continue to flow very slowly under the influence of gravity.

Supercritical fluid – a substance at a temperature above its critical temperature.

Supersaturated solution – contains a higher than saturation concentration of solute; slight disturbance or seeding causes crystallization of excess solute.

Suspension – a heterogeneous mixture in which solute-like particles settle out of the solvent-like phase sometime after their introduction. A mixture of a liquid and a finely divided insoluble solid.

T

Talc – a mineral representing the Mohs Scale, composed of hydrated magnesium silicate with the chemical formula $H_2Mg_3(SiO_3)_4$ or $Mg_3Si_4O_{10}(OH)_2$.

Temperature – a measure of heat intensity (i.e., hotness or coldness of a sample); a measure of kinetic energy of an object. Units are measured in degrees, and scales include Celsius, Fahrenheit, and Kelvin.

Temporary hardness – hardness in water, removed by boiling; caused by calcium hydrogen carbonate.

Ternary acid – a ternary compound containing H, O and another element, often a nonmetal.

Ternary compound – a compound consisting of three elements; may be ionic or covalent.

Tetrahedral – a term used to describe molecules and polyatomic ions with one atom in the center and four atoms at the corners of a tetrahedron.

Theoretical yield – the maximum amount of a specified product that could be obtained from specified amounts of reactants, assuming complete consumption of the limiting reactant according to only one reaction and complete recovery of the product; see *actual yield*.

Theory – a model describing the nature of a phenomenon.

Thermal conductivity – a property of a material to conduct heat (often noted as k).

Thermal cracking – decomposition by heating a substance in the presence of a catalyst and without air.

Thermochemistry – the study of absorption/release of heat within a chemical reaction; studies heat energy associated with chemical reactions and physical transformations.

Thermodynamics – studying the effects of changing temperature, volume, or pressure (or work, heat, and energy) on a macroscopic scale.

Thermodynamic stability – when a system is in its lowest energy state with its environment (equilibrium).

Thermometer – a device that measures the average energy of a system.

Thermonuclear energy – energy from nuclear fusion reactions.

Third Law of Thermodynamics – entropy of a pure crystalline substance at absolute zero equals zero.

Titration – a procedure in which one solution is added to another solution until the chemical reaction between the two solutions is complete; the concentration of one solution is known, and that of the other is unknown. The process of adding one solution to a measured amount of another to find out exactly how much of each is required to react.

Torr – a unit to measure pressure; 1 Torr is equivalent to 133.322 Pa or 1.3158×10^{-3} atm.

Total ionic equation – the expression for a chemical reaction written to show the predominant form of species in aqueous solution or contact with water.

Transition elements (metals) – B Group elements except IIB in the periodic table; sometimes called transition elements, elements with incomplete d sub-shells; the d-block elements.

Transition state theory –reactants pass through high-energy transition states before forming products.

Transuranic element – an atomic number greater than 92; none of the transuranic elements are stable.

Triple bond – the sharing of three pairs of electrons within a covalent bond (e.g., N_2).

Triple point – where the temperature and pressure of three phases are the same; water has a unique phase diagram.

Tyndall effect – results from light scattering by colloidal particles (a mixture where one substance is dispersed evenly throughout another) or by suspended particles.

U

Uncertainty –any measurement that involves estimating any amount that cannot be precisely reproduced.

Uncertainty principle – knowing the location of a particle makes the momentum uncertain, while knowing the momentum of a particle makes the location uncertain.

Unit cell – the smallest repeating unit of a lattice.

Unit factor – statements used in converting between units.

Universal (or ideal) gas constant – proportionality constant in the ideal gas law (0.08206 L·atm/(K·mol)).

UN number – a four-digit code used to note hazardous and flammable substances.

Unsaturated hydrocarbons – hydrocarbons that contain double or triple carbon-carbon bonds.

V

Valence bond theory – proposes that covalent bonds are formed when atomic orbitals on different atoms overlap and the electrons are shared.

Valence electrons – outermost electrons of atoms; usually those involved in bonding.

Valence shell electron pair repulsion theory (VSEPR) – assumes electron pairs are arranged around the central element of a molecule or polyatomic ion with maximum separation (and minimum repulsion) among regions of high electron density.

Valency – the number of electrons an atom wants to gain, lose, or share to have a full outer shell.

Van der Waals' equation – a quantitative relationship of a state extending the ideal gas law to real gases by including two empirically determined parameters, which are specific for different gases.

Van der Waals force – one of the forces (attraction/repulsion) between molecules.

Van't Hoff factor – the ratio of moles of particles in solution to moles of solute dissolved.

Vapor – when a substance is below the critical temperature in the gas phase.

Vaporization – the phase change from liquid to gas.

Vapor pressure – the particle pressure of vapor at the surface of its parent liquid.

Viscosity – the resistance of a liquid to flow (e.g., oil has a higher viscosity than water).

Volt – one joule of work per coulomb; the unit of electrical potential transferred.

Voltage – the potential difference between two electrodes; a measure of the chemical potential for a redox reaction.

Voltaic cells – electrochemical cells in which spontaneous chemical reactions produce electricity; known as *galvanic cells*.

Voltmeter – an instrument that measures the cell potential.

Volumetric analysis – measuring the volume of a solution (of known concentration) to determine the substance's concentration within the solution; see *titration*.

W

Water equivalent – the amount of water absorbing the same heat as the calorimeter per degree of temperature increase.

Weak electrolyte – a substance that conducts electricity poorly in a dilute aqueous solution.

Weak field ligand – a ligand that exerts a weak crystal or ligand field and generally forms high-spin complexes with metals.

X

X-ray – electromagnetic radiation between gamma and UV rays.

X-ray diffraction – a method for establishing structures of crystalline solids using single-wavelength X-rays and studying the diffraction pattern.

X-ray photoelectron spectroscopy – a spectroscopic technique used to measure the composition of a material.

Y

Yield – the amount of product produced during a chemical reaction.

Z

Zone melting – remove impurities from an element by melting and slowly traveling it down an ingot (cast).

Zone refining – a method of purifying a metal bar by passing it through an induction heater; this causes impurities to move along a melted portion.

Zwitterion (formerly called a dipolar ion) – a neutral molecule with a positive and negative electrical charge; multiple positive and negative charges can be present, distinct from dipoles at different locations within that molecule; known as *inner salts*.

Frank J. Addivinola, Ph.D.

This study guide's lead author and chief editor is Dr. Frank Addivinola. With his outstanding education, laboratory research, and decades of university science teaching, Dr. Addivinola lent his expertise to oversee the development of this series.

During his extensive career, Dr. Addivinola held faculty positions at colleges and universities, including Harvard University, Johns Hopkins University, University of Maryland, and Northeastern University, and taught undergraduate and graduate-level courses in biology, biochemistry, organic chemistry, inorganic chemistry, physics, anatomy and physiology, medical terminology, nutrition, and medical ethics. He received several awards for his research and presentations.

Dr. Frank Addivinola conducted original research in developmental biology as a doctoral candidate and pre-IRTA fellow in Molecular and Cell Biology at the National Institutes of Health (NIH). His dissertation advisor was Nobel laureate Marshall W. Nirenberg, Chief of the Biochemical Genetics Laboratory at the National Heart, Lung, and Blood Institute (NHLBI). Before NIH, Dr. Addivinola researched prostate cancer in the Cell Growth and Regulation Laboratory of Dr. Arthur Pardee at the Dana Farber Cancer Institute of Harvard Medical School.

Dr. Addivinola holds an undergraduate degree in biology from Williams College. He completed his Masters at Harvard University, Masters in Biotechnology at Johns Hopkins University, and five other graduate degrees at the University of Maryland University College, Suffolk University, and Northeastern University.

For online practice resources visit

https://www.sterling-prep.com

If you benefited from this book, please leave a review on Amazon so others

can learn from your input. Reviews help us understand our customers'

needs and experiences while keeping our commitment to quality.